PENGUIN BOOKS

Power Systems

Noam Chomsky is the author of numerous best-selling political books, including *Hegemony or Survival, Failed States, Interventions, What We Say Goes, Hopes and Prospects* and, most recently, *Occupy*, all of which are published by Hamish Hamilton/Penguin. He is a professor in the Department of Linguistics and Philosophy at MIT, and is widely credited with having revolutionized modern linguistics.

David Barsamian is the award-winning founder and director of Alternative Radio. He has authored several books of interviews with leading political thinkers.

www.chomsky.info
www.alternativeradio.org

ALSO BY NOAM CHOMSKY

Hegemony or Survival

Failed States

Imperial Ambitions

What We Say Goes

NOAM CHOMSKY

POWER SYSTEMS

CONVERSATIONS WITH
DAVID BARSAMIAN ON
GLOBAL DEMOCRATIC UPRISINGS
AND THE NEW CHALLENGES
TO US EMPIRE

PENGUIN BOOKS

PENGUIN BOOKS

Published by the Penguin Group
Penguin Books Ltd, 80 Strand, London WC2R 0RL, England
Penguin Group (USA) Inc., 375 Hudson Street, New York, New York 10014, USA
Penguin Group (Canada), 90 Eglinton Avenue East, Suite 700, Toronto, Ontario, Canada M4P 2Y3
(a division of Pearson Penguin Canada Inc.)
Penguin Ireland, 25 St Stephen's Green, Dublin 2, Ireland (a division of Penguin Books Ltd)
Penguin Group (Australia), 707 Collins Street, Melbourne, Victoria 3008, Australia
(a division of Pearson Australia Group Pty Ltd)
Penguin Books India Pvt Ltd, 11 Community Centre, Panchsheel Park, New Delhi – 110 017, India
Penguin Group (NZ), 67 Apollo Drive, Rosedale, Auckland 0632, New Zealand
(a division of Pearson New Zealand Ltd)
Penguin Books (South Africa) (Pty) Ltd, Block D, Rosebank Office Park,
181 Jan Smuts Avenue, Parktown North, Gauteng 2193, South Africa

Penguin Books Ltd, Registered Offices: 80 Strand, London WC2R 0RL, England

www.penguin.com

First published in the United States of America by Metropolitan Books 2012
First published in Great Britain by Hamish Hamilton 2013
Published in Penguin Books 2014
005

Printed in Great Britain by Clays Ltd, Elcograf S.p.A.

ISBN: 978–0–241–96524–5

www.greenpenguin.co.uk

CONTENTS

POWER SYSTEMS

1

THE NEW AMERICAN IMPERIALISM

Cambridge, Massachusetts (April 2, 2010)

One of the themes that Howard Zinn tried to address during his long career was the lack of historical memory. The facts of history are scrupulously ignored and/or distorted. I was wondering if you could comment on imperialism then and now, interventions then and now. Specifically about Saigon in 1963 and 1964 and Kabul today?

What happened in Vietnam in the early 1960s is gone from history. It was barely discussed at the time, and it's essentially disappeared. In 1954, there was a peace settlement between the United States and Vietnam. The United States regarded it as a disaster, refused to permit it to go forward, and established a client state in the South, which was a typical client state, carrying out torture, brutality, murders. By about 1960, the South Vietnamese government had probably killed seventy or eighty thousand people.[1] The repression was so harsh that it stimulated an internal rebellion, which was not what the North Vietnamese wanted. They wanted some time to develop their own society. But they were sort of coerced by the southern resistance into at least giving it verbal support.

By the time John F. Kennedy became involved in 1961, the situation was out of control. So Kennedy simply invaded the country. In 1962, he sent the U.S. Air Force to start bombing South Vietnam, using planes with South Vietnamese markings. Kennedy authorized the use of napalm, chemical warfare, to destroy the ground cover and crops. He started the process of driving the rural population into what were called "strategic hamlets," essentially concentration camps, where people were surrounded by barbed wire, supposedly to protect them from the guerillas who the U.S. government knew perfectly well they supported. This "pacification" ultimately drove millions of people out of the countryside while destroying large parts of it. Kennedy also began operations against North Vietnam on a small scale. That was 1962.

In 1963, the Kennedy administration got wind of the fact that the government of Ngo Dinh Diem it had installed in South Vietnam was trying to arrange negotiations with the North. Diem and his brother, Ngo Dinh Nhu, were trying to negotiate a peace settlement. So the Kennedy liberals determined that they had to be thrown out. The Kennedy administration organized a coup in which the two brothers were killed and they put in their own guy, meanwhile escalating the war. Then came the assassination of President Kennedy. Contrary to a lot of mythology, Kennedy was one of the hawks in the administration to the very last minute. He did agree to proposals for withdrawal from Vietnam, because he knew the war was very unpopular here, but always with the condition of withdrawal after victory. Once we get victory, we can withdraw and let the client regime go.

Actually, *imperialism* is an interesting term. The United States was founded as an empire. George Washington wrote in 1783 that "the gradual extension of our settlements will as certainly cause the savage, as the wolf, to retire; both being beasts of prey, tho' they differ in shape." Thomas Jefferson predicted that the "backward" tribes at the borders "will relapse into barbarism and misery, lose numbers by war and want, and we shall be obliged to drive them, with the beasts of the forests into the Stony mountains."[2] Once we don't need slavery anymore, we'll send the slaves back to Africa. And get rid of the Latins because they are an inferior race. We're the superior race of Anglo-Saxons. It's only to the benefit of everyone if we people the entire hemisphere.

But none of that is considered imperialism because of what some historians of imperialism call the "saltwater fallacy": it's only imperialism if you cross saltwater.[3] So, for example, if the Mississippi had been as wide as the

Irish Sea, let's say, then it would have been imperialism. But it was understood to be imperialism at the time—and it is. Settler colonialism, which is what this is, is by far the worst kind of imperialism, because it gets rid of the native population. Other kinds of imperialism exploit them, but settler colonialism eliminates them, "exterminates" them, to use the words of the Founding Fathers.

When the United States reached the geographic limits of what we call the national territory, U.S. expansionism just continued. Immediately. Eighteen ninety-eight, that's the year when the United States essentially conquered Cuba. The U.S. takeover was called "liberating" Cuba. In fact, Washington was preventing Cuba from liberating itself from Spain. Then the United States stole Hawaii from its population and invaded the Philippines. In the Philippines, U.S. troops killed a couple hundred thousand people, establishing a colonial system, which still exists.[4] That's one of the reasons why the Philippines has not joined the rest of East and Southeast Asia in the economic development of the past twenty or thirty years. It's an outlier. Part of the reason is it still retains the structure of the neocolonial system that the United States established.

But the new American imperialism seems to be substantially different from the older variety in that the United States is a declining economic power and is therefore seeing its political power and influence wane. I'm thinking, for example, of a Latin American hemisphere-wide organization that was recently formed that excludes the United States. Such a thing would have been unthinkable in the more than century-long U.S. domination of the continent.

I think talk about American decline should be taken with a grain of salt. The Second World War is when the United States really became a global power. It had been the biggest economy in the world by far for long before the war, but it was a regional power in a way. It controlled the Western Hemisphere and had made some forays into the Pacific. But the British were the world power. The Second World War changed that. The United States became the dominant world power. The wealth of the United States at that time is hard to believe. The United States had half the world's wealth. The other industrial societies were weakened or destroyed. The United States was in an incredible position of security. It controlled the hemisphere, both oceans, the opposite sides of both oceans, with a huge military force.

Of course, that declined. Europe and Japan recovered, and decolonization took place. By 1970, the United States was down, if you want to call it that, to about 25 percent of the world's wealth—roughly what it had been, say, in the 1920s. It remained the overwhelming global power, but not like it had been in 1950. Since 1970, it's been pretty stable, though of course there were changes.

I think what has happened in Latin America is not related to changes in the United States. Within the last decade, for the first time in five hundred years, since the Spanish and Portuguese conquest, Latin America has begun to deal with some of its problems. It's begun to integrate.[5] The countries were very separated from one another. Each one was oriented separately toward the West, first Europe and then the United States. That integration is important. It means that it's not so easy to pick the countries off one by one. We've actually seen that in crucial cases recently. Latin American nations can unify in defense against an outside force.

The other development, which is more significant and much more difficult, is that the countries of Latin America are beginning individually to face their massive internal problems. Latin America is just a scandal. With its resources, Latin America ought to be a rich continent, South America particularly. Almost a century ago, Brazil was expected to be the "colossus of the south," comparable to the United States, the so-called colossus of the north. In fact, Latin America has terrible poverty and extreme inequality, some of the worst in the world. Latin America has a huge amount of wealth, but it is very highly concentrated in a small—usually Europeanized, often white—elite, and exists alongside massive poverty and misery. There are some attempts to begin to deal with that, which is important—another form of integration—and Latin America is somewhat separating itself from U.S. control.

But the United States is reacting. In 2008, the United States was kicked out of its last military base in South

America, the Manta Air Base in Ecuador.[6] But it immediately picked up seven new military bases in Colombia, the one country that's still within the U.S. orbit—though so far the Constitutional Court has not granted the United States access to them.[7] President Barack Obama has added a couple more, as well as two naval bases in Panama.[8] In 2008, the Bush II administration reactivated the Fourth Fleet, the naval fleet that covers the Caribbean and Latin American waters, which had been deactivated in 1950, after the Second World War.[9] Government spending on training of Latin American officers is way up.[10] They're being trained to deal with what's sometimes called "radical populism."[11] That has a definite meaning in Latin America, and not a pretty one.

We don't have internal records, but it's very likely that Obama's support for the government installed by a military coup in Honduras—support not shared by Europe and Latin America—is related to the U.S. air base in the country.[12] Called the "unsinkable aircraft carrier" in the 1980s, the base was used for attacking Nicaragua and is still a major military base.[13] In fact, shortly after the military coup government took over, its leaders made a security deal with Colombia, the other U.S. client in the region.[14]

There are plenty of other complicated things happening in the world. There's a lot of talk about a global shift of power: India and China are going to become the new great powers, the wealthiest powers. Again, one should be pretty reserved about that. For example, there is a lot of talk about the U.S. debt and the fact that China holds so much of it. Actually, Japan holds more U.S. debt than China.[15] There have been occasions when China passed Japan, but most of the time, including right now, Japan holds most of the debt. When you put them together, the sovereign wealth funds of the United Arab Emirates probably hold more debt than China.[16]

Furthermore, the whole framework for the discussion of U.S. decline is misleading. We're taught to talk about the world as a world of states conceived as unified, coherent entities. If you study international relations (IR) theory, there's what's called "realist" IR theory, which says there is an anarchic world of states and states pursue their "national interest." It's in large part mythology. There are a few common interests, like we don't want to be destroyed. But, for the most part, people within a nation have very different interests. The interests of the CEO of General Electric and the janitor who cleans his floor are not the same. Part of the doctrinal system in the United States is the pretense that we're all a happy family, there are no class divisions, and everybody is working together in harmony. But that's radically false.

Furthermore, it's known to be false. At least, it has been for a long time. Take a dangerous radical like, say, Adam Smith, whom people worship but don't read. He said that in England the people who own the society make policy. The people who own the place are the "merchants and manufacturers." They're "the principal architects" of policy, and they carry it out in their own interests, no matter how harmful the effects on the people of England, which is not their business.[17] Of course, he was an old-fashioned conservative, so he had moral values. He was concerned with what he called the "savage injustice" of the Europeans, particularly what Britain was doing in India, causing famines and so on.[18] That's old-fashioned conservatism, not what's called conservatism now.

Power is no longer in the hands of the "merchants and manufacturers," but of financial institutions and multinationals. The result is the same. And these institutions have an interest in Chinese development. So if you're, say, the CEO of Walmart or Dell or Hewlett-Packard, you're perfectly happy to have very cheap labor in China working under hideous conditions and with no environmental constraints. As long as China has what's called economic growth, that's fine.

Actually, China's economic growth is a bit of a myth. China is largely an assembly plant. China is a major exporter, but while the U.S. trade deficit with China has gone up, the trade deficit with Japan, Singapore, and Korea has gone down. The reason is that a regional production system is developing. The more advanced countries of the region, Japan, Singapore, South Korea, Taiwan, send advanced technology, parts, and components to China, which uses its cheap labor force to assemble goods and send them out of the country. And U.S.

corporations do the same thing: they send parts and components to China, where people assemble and export the final products. Within the doctrinal framework, these are called Chinese exports, but they're regional exports in many instances and in other instances it's actually a case of the United States exporting to itself.

Once we break out of the framework of national states as unified entities with no internal divisions within them, we can see that there is a global shift of power, but it's from the global workforce to the owners of the world: transnational capital, global financial institutions. So, for example, the earnings of working people as a percentage of national income has by and large declined in the last couple of decades, but apparently it's declined in China more than in most places.[19] There is certainly economic growth in China and India. Hundreds of millions of people live a lot better than they did before, but then there are hundreds of millions more who don't. In fact, it's getting worse for them in many ways.[20]

The UN Human Development Index ranks India as 134th, slightly above Cambodia and Laos. And China ranks 101st.[21]

India is about where it was twenty years ago, before the famous reforms began. So yes, there has been growth. You go to Delhi, there is plenty of wealth. But it's an extension of the traditional Third World system. Even in the worst days, if you went to the poorest country in the world, say,

Haiti, you would find a sector—white, European, mulatto maybe—that lives in tremendous wealth and luxury. You find the same structure in India, just on a vastly different scale. So in India a couple hundred million people now have cars, television sets, and nice homes. You have multi-billionaires in India who are building palaces for themselves.[22] Meanwhile, the consumption of food, on average, has actually declined during this period of growth.[23]

Incidentally, the richest man in the world is now from Mexico, Carlos Slim. He beat Bill Gates this year.[24] As one of the consequences of privatization in Mexico, mainly over the last twenty to thirty years, he was given a telecommunications monopoly.

I think you have to take the ranking of China with a grain of salt. India is a much more open society, so we know a lot more about what is happening there. China is pretty closed. We don't know much about what's going on in China's rural areas. One person doing important research on this is Ching Kwan Lee, a sociologist at UCLA. She's done extensive study of Chinese labor conditions, and she distinguishes what she calls the rustbelt from the sunbelt.[25] The rustbelt is up in the northeast, the big production center where the state-run industrial sector was based. That's being wiped out. And she compares it to the rustbelt in the United States. Workers have essentially nothing. They had a compact, they thought. People have done studies of workers in Ohio and Indiana. They feel cheated, rightly. They thought they had a deal with the corporations and the government: they would work hard all their lives and in return they would get pensions, they would get security, their children would get jobs. They served in the army, they did all the right things. Now they're being thrown into the trash can. No pensions, no security, no jobs. The jobs are being shipped somewhere else. She finds the same in the Chinese rustbelt, except there the compact was the Maoist version: we have solidarity, we build the country, we sacrifice, and then we get security.

The sunbelt is southeast China, the big production center now, where the factories are bringing younger workers in from the rural areas. These workers don't have this Maoist tradition of solidarity and working to build the country. They're peasants. In fact, their lives are still based in the villages. That's where their families are, where they raise children, where they can go if they lose their job. They're a migrant workforce.

There is huge labor unrest all over China. In the southeast, the sunbelt, it's because the government is failing its legal obligations. There are laws that say you should have certain wages and working conditions, but workers don't have anything. So they're protesting that. There are a huge number of protests, even by official statistics.[26] The labor force is atomized but very militant. But we really don't know what's going on in the inner rural areas. On top of which, there are enormous ecological problems developing in China.

So if you measured growth rationally—counting not

just the number of products you make but the costs and benefits of making them—China's growth rate would be much lower. And its ranking in the Human Development Index would probably also be lower, though 101st is bad enough.

On your office door at MIT you have a bumper sticker featuring a quote from the two-time Medal of Honor winner Major General Smedley Butler, who was a veteran of many U.S. interventions, from China to Nicaragua. The sticker says, "War is a racket. The few profit, the many pay."

In fact, he very eloquently described the way war is a racket. He says, "I was a racketeer for capitalism," and he describes his role in many interventions.[27] Actually, a very timely example is Haiti. When Woodrow Wilson invaded Haiti in 1915, Smedley Butler was one of the commanders, though not the top one. He was the person who President Woodrow Wilson sent to disband the parliament. The parliament of Haiti had refused to accept a U.S.-written constitution, which permitted American corporations to buy up Haitian land. This measure was considered very progressive. If you go back to the time, the big thinkers were saying that Haiti needs foreign investment in order to develop. You can't expect American investors to put money in Haiti unless they can own the place, so we have to have this progressive legislation. And these backward people don't understand it, so we have to disband the parliament. Butler says we disbanded them by typical Marine Corps measures, at gunpoint. After that, the marines, under Butler, ran a referendum in which they got 99.9 percent approval of the U.S.-imposed constitution, with 5 percent of the population participating—namely, the rich elite.[28] That was considered a great democratic achievement. It was another step in the process of driving the population off the land, turning them into assembly plant workers or something considered to be to their "comparative advantage" by progressive thinkers. And finally you get the hideous catastrophe we've just seen with the January 2010 earthquake in Haiti.

In his later years, Butler was pretty bitter. He also stopped a business coup that planned to overthrow the administration of and kill President Franklin D. Roosevelt.[29] He intervened and somehow put an end to it. He was vilified for speaking out, but he was a real hero.

Let's talk more about Afghanistan and the U.S. war there. In March 2010, Obama visited Bagram air base.[30] It is a site of major war crimes, which went virtually unmentioned in news reports. Obama told the troops that their mission was "absolutely essential," declaring, "We did not choose this war. This was not an act of America wanting to expand its influence; of us wanting to meddle in somebody else's business. We were attacked viciously on 9/11." And finally he told the assembled troops, "If I thought for a minute that America's vital interests were not served, were not at stake here in Afghanistan, I would order all of you home right away."[31] What are those vital interests from Obama's point of view?

There are a few strategic interests but, by this point, I suspect it's mostly domestic politics. Daniel Ellsberg observed this about the war against Vietnam. If you pull out without victory, which is called losing, you're literally dead. Obama inherited the war. And I suspect the dominant interest is self-preservation.

The United States didn't invade Afghanistan because we were viciously attacked. It's true that there was an attack on 9/11, but the government didn't know who did it. In fact, eight months later, after the most intensive international investigation in history, the head of the Federal Bureau of Investigation informed the press that they still didn't know who did it. He said they had suspicions. The suspicions were that the plot was hatched in Afghanistan but implemented in Germany and the United Arab Emirates, and, of course, in the United States.[32]

After 9/11, Bush II essentially ordered the Taliban to hand over Osama bin Laden, and they temporized. They might have handed him over, actually. They asked for evidence that he was involved in the attacks of 9/11. And, of course, the government, first of all, couldn't give them any evidence because they didn't have any. But, secondly, they reacted with total contempt. How can you ask us for evidence if we want you to hand somebody over? What lèse-majesté is this? So Bush simply informed the people of Afghanistan that we're going to bomb you until the Taliban hand over Osama bin Laden. He said nothing about overthrowing the Taliban. That came three weeks later, when British admiral Michael Boyce, the head of the British Defense Staff, announced to the Afghans that we're going to continue bombing you until you overthrow your government.[33] This fits the definition of terrorism exactly, but it's much worse. It's aggression.

How did the Afghans feel about it? We actually don't know. There were leading Afghan anti-Taliban activists who were bitterly opposed to the bombing. In fact, a couple of weeks after the bombing started, the U.S. favorite, Abdul Haq, considered a great martyr in Afghanistan, was interviewed about this. He said that the Americans are carrying out the bombing only because they want to show their muscle. They're undermining our efforts to overthrow the Taliban from within, which we can do. If, instead of killing innocent Afghans, they help us, that's what will happen.[34] Soon after that, there was a meeting in Peshawar in Pakistan of a thousand tribal leaders, some from Afghanistan who trekked across the border, some from Pakistan. They disagreed on a lot of things, but they were unanimous on one thing: stop the bombing.[35] That was after about a month. Could the Taliban have been overthrown from within? It's very likely. There were strong anti-Taliban forces. But the United States didn't want that. It wanted to invade and conquer Afghanistan and impose its own rule.

The same was true in Iraq. If it hadn't been for the

sanctions, it's very likely that Saddam Hussein would have been overthrown from within in much the same way as a whole rogues' gallery of other gangsters the United States and Britain have supported, like, say, Nicolae Ceauşescu, the worst of the Eastern European dictators. Nobody wants to talk about him anymore, but the United States supported him until the very end. Suharto in Indonesia, Ferdinand Marcos in the Philippines, Jean-Claude Duvalier in Haiti, Chun Doo-hwan in South Korea, Mobutu Sese Seko in Zaire—they were all overthrown from within. But the United States didn't want that in Iraq. It wanted to impose its own regime. And the same in Afghanistan.

There are geostrategic reasons. They're not small. How dominant they are in the thinking of planners we can only speculate. But there is a reason why everybody has been invading Afghanistan since Alexander the Great. The country is in a highly strategic position relative to Central Asia, South Asia, and the Middle East. There are specific reasons in the present case having to do with pipeline projects, which are in the background. We don't know how important these considerations are, but since the 1990s the United States has been trying hard to establish the Trans-Afghanistan Pipeline (TAPI) from Turkmenistan, which has a huge amount of natural gas, to India. It has to go through Kandahar, in fact. So Turkmenistan, Afghanistan, Pakistan, and India are all involved.

The United States wants the pipeline for two reasons. One reason is to try to prevent Russia from having control of natural gas. That's the new "great game": Who controls Central Asian resources? The other reason has to do with isolating Iran. The natural way to get the energy resources India needs is from Iran, a pipeline right from Iran to Pakistan to India. The United States wants to block this from happening in the worst way. It's a complicated business. Pakistan has just agreed to let the pipeline run from Iran to Pakistan.[36] The question is whether India will try to join in. The TAPI pipeline would be a good weapon to try to undercut that.

In fact, that's probably one of the main reasons why the United States entered into a deal with India in 2008 to permit India to openly violate the Non-Proliferation Treaty and to import nuclear technology—which, of course, can be transferred to weapons production.[37] That's another way to try to draw India more into the U.S. orbit and separate it from Iran.

So all of these things are going on. There are a lot of broad considerations involved. But I still suspect that domestic politics is uppermost. We can't get out of Afghanistan without victory or we'll be slaughtered.

Is that related to the greatly expanding drone attacks on Pakistan?

Yes. They're horrible, but they're also interesting. They tell us a lot about American ideology. The drone attacks are not a secret. There's much we don't know about them, but mostly they're not a secret. The Pakistani population is overwhelmingly opposed to them, but they're justified here on the grounds that the Pakistani leadership covertly agrees.[38] Fortunately for us, Pakistan is so dictatorial that they don't have to pay much attention to their population.[39] So if the country is a brutal dictatorship, it's great, because the leaders can secretly agree to what we're doing and disregard their population, which is overwhelmingly opposed to it. Pakistan's lack of democracy is considered a good thing. And then in an adjacent newspaper article you read, "We're promoting democracy." It's what George Orwell called "doublethink," the ability to have two contradictory ideas in mind and believe both of them.[40] That's almost a definition of our intellectual culture. And this is a perfect example of it. Yes, the bombing is fine, because secretly the leadership agrees, even though they have to tell the population they're against it because the population is overwhelmingly opposed.

India, Pakistan's neighbor, has seen a huge surge in internal resistance to neoliberalism. Manmohan Singh, the current prime minister, was the finance minister in the early 1990s. He let the cat out of the bag when he told the Indian parliament in June 2009, "If left-wing extremism"—the catchall phrase for Naxalites, Maoists, terrorists—"continues to flourish in important parts of our country which have tremendous natural resources of minerals and other precious things, that will certainly affect the climate for investment."[41]

It's certainly true. There are foreign investors and, for that matter, Indian investors who want to get into these resource-rich areas, even if that means, of course, getting rid of the tribal people, destroying their way of life. But India has been at war internally ever since its founding. In fact, this war goes back way before, to the British in earlier periods. Large parts of India are at war at the moment. Whole states are under attack. You have to get the resources for what's called economic growth.

India figures into U.S. geostrategic planning vis-à-vis China. There has been a major expansion of U.S. weapon sales to India, training, intelligence sharing.[42] *Israel is involved, as well.*[43] *How has India gone from a country that was once nonaligned to one that's become very aligned with Washington?*

India was not only nonaligned, it was a leader of the nonaligned movement. It had pretty close military relations with Russia, but in both power and ideology it was at the core of the nonaligned movement. It's shifted. India is playing a complicated game. It's keeping its relations with China, although there are also conflicts with China. So economic and other relations with China are proceeding. At the same time, there is a conflict with China in the Ladakh area. The Sino-Indian War was fought there in 1962, and it still remains a conflicted area.

I think India is trying to decide how to position itself in the global system. The relations with the United States and with Israel, its U.S. client, are very close. Indian forces attacking the tribal areas are apparently using Israeli technology.[44] For years, one of the services Israel has provided to the United States is to carry out state terrorism. It's very efficient at doing that. Israelis did it in South Africa and Central America.[45] Now they're doing it in India. They're probably doing it in Kashmir—it's claimed, but I don't know if it's true—and very likely in the Kurdish areas in northern Iraq.[46]

Israel has been a hired gun for thirty years and has helped the United States—by "the United States," I mean the White House—get around congressional sanctions. For example, there were congressional sanctions against giving aid to Guatemala, the worst of the terrorist states of Central America. So Washington funneled money through Israel and Taiwan.[47]

The United States is a big power. Small countries hire individual terrorists like Carlos the Jackal. The United States hires terrorist states. It's much more efficient. You can do a much more murderous and brutal job. Israel is one. Taiwan is another. Britain has also played that role.

Indian-Israeli relations have gotten very close as part of the overarching U.S. effort to maintain a global system that will give the United States a geostrategic advantage over China. But it's complex. China, for example, is now moving into Saudi Arabia, the real heartland of U.S. concerns. I think China may be the leading importer by now of Saudi oil.[48] And China has had a historic relationship with Pakistan. It's now moving to develop a port system in Karachi and Gwadar, which would be a way for China to get access to the South Asian seas and also key for importing oil and even minerals from Africa.[49] Actually, the same thing is going on in Latin America. China is now probably the leading trading partner of Brazil. It has surpassed the United States and Europe.[50]

We were both at a talk that Arundhati Roy gave at Harvard describing the rather extraordinary amount of resistance to neoliberal policies in India.[51] There is a tremendous amount of push-back. I wrote to Howard Zinn about her talk. He wrote back to me, in one of the last e-mails I received from him, "Compared to India, the United States seems like a desert."

It wasn't at one time. If you go back to the nineteenth century, the indigenous population of the United States resisted. In this respect, the United States is a desert because we exterminated the native people. The United States won that war. By the end of the nineteenth century, the indigenous people were essentially gone. India is now in the stage the United States was in during the nineteenth century.

I'm thinking more of workers here who have lost their jobs, who have lost their pensions and benefits. At a talk you gave in Portland, Oregon, called "When Elites Fail," you decried the fact that the Left has not been able to mobilize dissent.[52] *The Right has certainly been able to.*

That's true. But I don't think India is a good comparison. Earlier periods in U.S. history are a better comparison.

Take, say, the 1930s. The Depression hit in 1929. About five years later, you started getting real militant labor organizing, the forming of the Congress of Industrial Organizations, sit-down strikes.[53] That's what basically impelled Roosevelt to carry out the New Deal reforms. That hasn't happened in the current economic crisis. Remember the 1920s were a period when labor was almost completely crushed. One of the leading labor historians in the United States, David Montgomery, has a book called *The Fall of the House of Labor*.[54] The rise of the house of labor was from the nineteenth-century militants on through the early-twentieth-century labor agitation that was crushed by Woodrow Wilson, who was as brutal internally as he was externally. The Red Scare almost decimated the workers' movement. That was the 1920s. There was a change in the 1930s, in the course of the Depression. But it took quite a few years. And the Depression was much worse than the current recession. This is bad enough, but that was much worse.

And then there were other factors. For example, we're not supposed to say it, but the Communist Party was an organized and persistent element. It didn't show up for a demonstration and then scatter so somebody else then had to start something else. It was always there—and it was in for the long haul. That's not the type of organization we have now. And the Communist Party was in the forefront of civil rights struggles, which were very significant in the 1930s, as well as labor organizing, union struggles, union militancy. They were a spark, which is lacking now.

Why is it lacking?

First of all, the Communist Party was totally crushed. In fact, the activist Left was crushed under President Harry S. Truman. What we call McCarthyism was actually started by Truman. The unions did grow in size, but they grew as collaborationist unions. That's one of the reasons why, say, Canada, a very similar country, has a health care system and we don't. In Canada, the unions struggled for health care for the country. In the United States, they struggled for health care for themselves. So if you're an autoworker here in the United States, you had a pretty good health care and pension system. Union workers won health care for themselves in a compact with the corporations. They thought it was a deal. What they couldn't see was that it's a suicide pact. If the corporation decides the compact is over, then it's over. Meanwhile, the rest of the country didn't get health care. So now the United States has a completely dysfunctional health care system, while Canada has one that more or less works. That's a reflection of different cultural values and institutional structures in two very similar countries. So yes, the working class did continue to develop and grow here, but with class collaboration, that is, in a compact with the corporations.

You may recall in 1979, Doug Fraser, who was the head of the United Auto Workers, gave a speech in which he lamented the fact that business was engaged in what he called "a one-sided class war" against working people.[55] We thought we were all cooperating. That was pretty dumb. Business is always engaged in a one-sided class war, especially in the United States, which has a very class-conscious business community. They're always militantly struggling to get rid of any interference with their domination and control. The labor unions went along with it. They benefited their own workers temporarily. Now they're paying the penalty.

In a lecture at the Left Forum in New York on March 21, 2010, you talked about Joseph Stack and his manifesto.[56] He is the man who took a plane and flew it into the IRS building in Austin.[57] You went on to talk about the Weimar Republic. You said, "All of this evokes memories of other days when the center did not hold, and they're worth thinking about." Talk about Stack. And why did you bring up Weimar?

Joe Stack left a manifesto, which liberal columnists just ridiculed. They dismissed him as a crazy person. But if you read his manifesto, it's an eloquent and insightful commentary on contemporary American society. He starts by describing how he grew up in an old industrial area. It happened to be Harrisburg, Pennsylvania. When he was about eighteen or nineteen, he was a college student living on a pittance. In his building there was an eighty-year-old woman who was living on cat food. And he tells her story. Her husband had been a steelworker, someone who belonged to what is called the "privileged working class," the part that made out pretty well during the period of economic growth in the 1950s and 1960s. He was guaranteed a pension. He looked forward to his retirement. It was all stolen from him. He died prematurely. That happens pretty commonly among people who are faced with these situations. His future was stolen by the company, by the government, and by the union. And she's left eating cat food. That was his first recognition that something was wrong with the picture of the world that he had been taught in grade school. Then he goes on to say, "I decided that I didn't trust big business to take care of me, and that I would take responsibility for my own future and myself."

He talks about his own efforts over the years to start a small business and how at every point he was smashed down by corporate power, by the government. Finally, he got to the point of saying, we've got to "revolt," and the only way to revolt is to awaken people from their torpor and show that we're willing to die for our freedom. And then he smashed himself into the building in Austin as a wake-up call to the many people like him.

So what's happening to what we call the middle class—because we're not allowed to use the word *working class*. That's what's happening to working people. In other countries, it's called the working class. But here everybody has to be middle class or underclass.

The Left Forum used the phrase "the center cannot hold" as the title of the conference at which I spoke, and correctly. What's happening all over the United States is tremendous anger against corporations, against the government, against the political parties, against institutions, against professions. About half the population thinks that every person in Congress, including their own representative, should be thrown out.[58] That's the center not holding.

Take a look back at the Weimar Republic. It's not a perfect analogy by any means, but it's strikingly similar. First of all, Germany was at the peak of Western civilization in the 1920s—in the arts, sciences, and literature. It was considered a model democracy. The political system was lively. There were large working-class organizations, a huge Social Democratic Party, a big Communist Party, many civic institutions. The country had plenty of problems but it was, by any standards we have, a vibrant democratic society.

Germany was beginning to change even before the Depression. In 1925, there was a mass popular vote for Paul von Hindenburg for president. He was a Prussian aristocrat, yet his supporters were petty bourgeois storekeepers, disillusioned workers, and others—in fact, demographically not unlike the Tea Party movement. And they became the mass base for Nazism. In 1928, the Nazis still got under 3 percent of the vote. In 1933—that's only five years later—they were so powerful that Hindenburg had to appoint Adolf Hitler as chancellor. Hindenburg hated Hitler. Again, Hindenburg was an aristocrat, a general. He didn't pal around with the hoi polloi. And Hitler was this "little corporal," as he called him. What the heck is he doing in our aristocratic Germany? But he had to appoint him as chancellor because of his mass base. That was within five years.

If you look at the forces behind this shift, initially one was disillusionment with the political system. The parties were bickering. They weren't doing anything for the people. By then, the Depression had hit and the Nazis could appeal to nationalism. Hitler was a charismatic leader. We're going to create a powerful new Germany, which is going to find its proper place in the sun. We have to fight our enemies: the Bolsheviks and the Jews. They're the trouble. That's what's spoiling Germany. By 1933, Hitler for the first time declared May Day a workers' holiday. The Social Democrats, who were a powerful group, had been trying to do that ever since the Second Reich was established, but they could never do it. Hitler did it. There were huge demonstrations in Berlin, which was called "Red Berlin," a working-class, left-wing city. There were about a million people demonstrating, very excited. Our new united Germany is going to forge a new way. End all this political nonsense by the parties, and we'll become a unified, organized, militarized country that can show the world what real power and authority is.

All of that looks very similar to here. It's ominous. The Nazis destroyed the major working-class organizations. The Social Democrats and the Communists were huge organizations, not just political parties. They had clubs, associations, and civic societies.[59] They were all wiped out, partly by force but partly because the people joined the Nazis out of disillusionment and hope for a better future, a bright militaristic, jingoist future. I wouldn't say it's identical, but the parallels are strong enough to be frightening. You can see someone like Joe Stack joining that group.

Arundhati Roy has decried weekend protesters. You go to a march or a demonstration and then back to the usual routine on Monday. She's said that it's necessary that risks be taken, that protests have consequences.

I'm not sure that I agree with her that risks are important. Of course, serious demonstrations are going to have risks. You can get arrested. But the real issue, I think, is continuity. Going home is the problem. That's why the old Communist Party was so significant. There was always somebody around to turn the mimeograph machine. They were in it for the long haul. They didn't expect quick victories. Maybe you win something, maybe you don't, but then you lay the basis for something else, you go on to the next thing. That mentality is basically missing here. And it was during the 1960s, too.

It was missing in the 1960s?

Yes. If you go back to the 1960s, the big demonstrations, like the Columbia student strike and the marches on Washington, an enormous number of the young people involved thought that they were going to win. If we sit in the president's office for three weeks, we're going to have love and peace in the world. You recall that, I'm sure. Of course, love and peace didn't happen, so they were disillusioned and gave up. That lack of continuity has to be overcome.

For a while it was overcome in the civil rights movement. Many people in the movement knew it would be a long struggle. We're not going to win right away. Maybe we will get something, but then we'll hit a barrier. They managed to keep going until they tried to expand the African American civil rights movement to become a poor people's movement. This was Martin Luther King's inspiration, which was to extend the civil rights movement.

So, just to take King, because he's visible. On Martin Luther King Day, he's greatly celebrated for what he did in the early 1960s when he was saying "I Have a Dream" and "let's get rid of racist sheriffs in Alabama." That was okay. But by 1965 he was getting to be a dangerous figure. For one thing, he was turning against the war in Vietnam pretty strongly. For another, he was working to be at the head of a developing poor people's movement. He was assassinated when he was taking part in a strike of sanitation workers and he was on his way to Washington for a poor people's convention. He was going beyond racist sheriffs in Alabama to northern racism, which is much more deep-seated and class-based. The civil rights movement was partly destroyed by force and partly frittered away at that stage. It never really made it past the point where you get into class issues.

What Arundhati said about not going home is the crucial part. You have to understand that you're not going to win by sitting in the president's office. You don't get a world of love and peace that way. You may get a little victory, but then you're going to have a bigger struggle ahead. It's like mountain climbing. You climb a peak, you think you're at the top—and then you notice there is a bigger peak right beyond it, and you've got to climb that one. That's what popular struggle is like. And that's lacking. Our quick-gratification culture is not conducive to that kind of commitment.

There are people and organizations that really are persistent and struggle—and, of course, those are the ones under attack. Take ACORN, the Association of Community Organizations for Reform Now. Why was ACORN destroyed? There was a little bit of a scam, but by the standards of corporate corruption what they did was trivial. But instantly the media, Congress, everybody jumped on the news and destroyed them.[60] Because it's a persistent organization working for poor people, and that's dangerous.

Given the dismal economic situation, why isn't there a left response? Certainly the Right has generated answers and explanations.

So did Hitler. It was the Jews and the Bolsheviks. They're crazy answers, but they are answers.

It's better than being in a vacuum. The Left seems to have nothing to say.

The Democratic Party and even the Democratic Left are not going to tell people, "Look, your problem is that, back in the 1970s, we took part in a major process of financialization of the economy and the hollowing out of the productive system. So your wages and income have stagnated for thirty years, while what wealth is produced is in a very few pockets. Those are our policies." They're not going to tell them that. No, there is no real Left now. If you are just counting heads, there are probably more people involved than in the 1960s, but they are atomized, committed to different special interests—gay rights, environmental rights, this, that. They don't coalesce into a movement that can really do things.

And there are things that could be done, which I talked about a little in the Left Forum lecture you mentioned. For example, the Obama administration essentially owns the auto industry at this point, except for Ford. Certainly GM. What they're doing is continuing the policies of closing down GM plants, which means destroying the workforce, destroying communities. The communities were built by the unions. Meanwhile, Obama sends emissaries to tell people in these cities, "We really care about you and want to help you," and distribute some pennies. At almost the same time, he sends another emissary, the secretary of transportation, to Spain to spend federal stimulus money for contracts with Spanish companies to build high-speed rail facilities.[61] Those high-speed rail facilities could be built in the factories that are being closed, but that's not important from the point of view of the bankers and Smith's "principal architects" of policy.

What's lacking is the consciousness that began to arise in the 1930s—we'll take it over and run it ourselves. The things that really put the fear of God into manufacturers and the government in the 1930s were the sit-down strikes. A sit-down strike is just one step short of saying, "Look, instead of sitting down, we'll run this place. We don't need owners and managers." That's huge. That could be done in Detroit and in other places that are being closed down.

2

CHAINS OF SUBMISSION AND SUBSERVIENCE

BOULDER, COLORADO (MARCH 31, 2011)

Formal slavery has long been abolished, but a de facto mental slavery has replaced it. This is reflected in obedience to power and authority. People are reduced to asking, pleading with the masters for favors, a few crumbs here and there. Don't slash the budget by this amount or don't cut this after-school program by this much. How does a person break the chains of submission and subservience?

First of all, mental slavery didn't replace slavery—it has always existed. How do you break mental slavery? There is no magic answer. You start by asking for reforms that make sense. You see if they come. If they do, you try to go farther. Or if you hit a brick wall, if the power systems won't relent, you move on to try to overthrow them. That's the history of activism. That's how slavery ended.

Is it more difficult to do here in the United States than, say, Bolivia?

I think it's a lot easier here than in Bolivia. Just as it's easier to protest here than it is in Tahrir Square in Egypt. Bolivians have far harsher circumstances. What they've achieved is remarkable. The circumstances are much harder, but it has to be done.

To what extent does the propaganda system induce docility and passivity in the citizenry in the United States?

That's its point. But that has been its point from time immemorial. It's part of the function of the reverence for kings, priests, submission to religious authorities. These are doctrinal characteristics of power systems that seek to induce passivity. The major propaganda systems that we face now, mostly growing out of the huge public relations industry, were developed quite consciously about a century ago in the freest countries in the world, in Britain and the United States, because of a very clear and articulated recognition that people had gained so many rights that it was hard to suppress them by force. So you had to try to control their attitudes and beliefs or divert them somehow. As the economist Paul Nystrom argued, you have to try to fabricate consumers and create wants so people will be trapped.[1] It's a common method.

It was used by the slave owners. For example, when Britain abolished slavery, it had plantations using slaves throughout the West Indies. With official slavery gone, there were big parliamentary debates about how to sustain the same regime. What would stop a former slave from going up into the hills, where there was plenty of land, and just living happily there? They hit on the same method that everyone hits on: try to capture them with consumer goods. So they offered teasers—easy terms, gifts. And then when people got trapped into wanting consumer goods and started getting into debt at company stores, pretty soon you had a restoration of something similar to slavery, from the plantation owners' point of view.[2]

The United Fruit Company independently did the same thing in Central America, and U.S. and British business communities independently hit on the same technique in the early twentieth century. Out of that developed this enormous propaganda system directed exactly as Nystrom said, toward fabricating consumers and "concentrating human attention on the more superficial things."[3] And, of course, it also goes along with trying to control people's ideas and beliefs. That's another part of the doctrinal system.

These techniques aren't new. They're as old as the

hills. But they take new forms as circumstances change. The techniques we now see are a reaction to the achievement of earlier generations in gaining a lot more freedom. And I must say, it's a lot easier to combat the fabrication of consumers than torture chambers.

As you travel around the United States, you've often commented that communities with community radio stations are marginally different from those without them. For example, your hometown of Boston does not have a community radio station.

It's not a scientific conclusion—it's an impression—but, yes, Boston is a good example. There is no community radio station, and things are very scattered. People don't know that something is happening in another part of town. There is no interaction, no way to get people together. There are other means, such as the Internet, but there is no place you can turn to directly to find out what's happening, even to gain a critical analysis of what's going on in the world that relates to local concerns and interests. And that does hamper the ability to create community.

You're an educator. You've taught at MIT for decades. Many people are concerned about what's happening to public education. There are announcements of layoffs of thousands of teachers all over the country, larger class sizes, closing of schools, huge budget cuts. Remedial programs are being reduced or eliminated altogether. Don't the powers that be, the corporate elites, need a trained and competent workforce? Or will they rely simply on South and East Asians for that?

I don't think the business world, at least in the short term, is that concerned about lacking a workforce. First of all, there has been a substantial program of offshoring of production for the last thirty years. Not just manual labor, but also data analysis. You have a much cheaper workforce abroad. In fact, a couple of years ago, IBM announced inducements to try to get its U.S. staff to move to India, where they could live on smaller salaries.[4] So what you said is partly true. But I think corporate elites assume they can maintain a big enough domestic workforce with a smaller part of the population.

The developments you describe are all part of a major effort to undermine public education altogether, basically to privatize it, which would be a big boon to private power. Private power doesn't like public education, for many reasons. One is the principle on which it's based, which is threatening to power. Public education is based on a principle of solidarity. So, for example, I had my children fifty years ago. Nevertheless, I feel and I'm supposed to feel that I should pay taxes so that the kids across the street can go to school. That's counter to the doctrine that you should just look after yourself and let everyone else fall by the wayside, a basic principle of business rule. Public education is a threat to that belief system because it builds up a sense of solidarity, community, mutual support.

The same is true of Social Security. That's one of the reasons that there is such a passionate attempt to destroy Social Security, even though there are no economic reasons to do so, none of any significance at least. But public education and Social Security are residues of a dangerous conception that we're all in this together and we have to work together to create a better life and a better future. If you're trying to maximize profit or maximize consumption, then working together is the wrong idea. It has to be beaten out of people's heads.

Solidarity makes people hard to control and prevents them from being passive objects of private power. So you have to have a propaganda system that overcomes any deviations from the principle of subjugation to power systems.

There are major efforts to replace public schools with semi-privatized systems that would still be supported by the public but run more or less privately, such as charter schools. There is no evidence that they're any better.[5] For all we know, they're even worse. But this privatization of schools does undermine solidarity and mutual support—dangerous ideas that harm concentrated power.

Certainly labor unions in the United States have historically been institutions of solidarity. From a peak of representing 35 percent of all workers, union membership is now down to single digits.[6] *Workers are being asked to work longer hours, their wages and benefits are being reduced, they're losing jobs. Is capital using the current economic crisis to implement its long-term project to smash unions?*

Unions are bitterly hated by private power. That's always been true. The United States is a business-run society, much more so than comparable ones. Correspondingly, it has a very brutal labor history, much worse than in other societies. There have been constant efforts to try to destroy unions. By the 1920s, for example, they were almost crushed. Then they came back again in the worker struggles in the 1930s. It took almost no time for the business world to organize to try to destroy them again. Immediately after the war, you had the Taft-Hartley bill, other antilabor measures, and immense propaganda campaigns—in the churches, schools, cinema, press—to turn people against unions.

Over time, this campaign had some success, but a majority of the workforce would still prefer to be unionized if they could be.[7] Barriers have been set up by state policy which make it very hard to join a union.[8] The consequence of all this is that private-sector unionization is down to about 7 percent.[9] Public-sector unions still haven't been destroyed, but that's why there is a bitter attack against them right now. The attack in Wisconsin on the right of workers to organize and collectively bargain is a clear example.[10] The issues in Wisconsin have nothing to do with the state budget deficit. That's a fraud that's simply used as a pretext. The issue is the right of collective bargaining, one of the basic principles of union organization. The business world wants to destroy that.

Rhetoric aside, has the Democratic Party really been a friend of organized labor and the working class?

Compared with the Republicans, yes, but that's not saying much. The studies of Larry Bartels and other political scientists show that working people and the poor tend to do somewhat better under Democratic than Republican administrations.[11] But that just means that the Republicans are deeper in the pockets of the corporate system than the Democrats are. They're both nuzzled there quite happily. There are individual members of the Democratic Party who have been friends of labor, but they're a scattered and diminishing minority.

Take, say, Obama. The lame-duck session of Congress after the midterm elections in November 2010 was interesting. He was highly praised, including by his supporters, for his statesmanlike attitude during the lame-duck session, bipartisanship, and getting legislation through.[12] What did he get through? The main achievement was a huge tax cut for the extremely wealthy.[13] When I say extremely wealthy, I mean extremely. So, for example, I'm pretty well off, but I was below the cutoff point for the tax break. That break was a huge gift to a tiny sector of the wealthy, the deficit even higher—but who cares about that? That was Obama's major achievement. Meanwhile, at the same time, he initiated a tax increase on federal workers. Of course, no one called it a tax increase. That doesn't sound good. They called it a pay freeze.[14] But a pay freeze on public-sector workers is exactly the same thing as a tax increase. So we punish public-sector workers and reward the executives of Goldman Sachs, who just announced a $17.5 billion compensation package for themselves.[15]

In a talk that you gave at the University of North Carolina at Chapel Hill called "Human Intelligence and the Environment," you said that the system "is just driving us to disaster."[16] I note that you said this many months before the political upheavals in Madison, Wisconsin. "So is anything going to be done about it?" you asked. And answered, "The prospects are not very auspicious." Why not?

The prospects are not auspicious because of the general feeling you described before that there is nothing we can do. As long as people sit by passively and let things happen to them, the dynamics of the system will drive it in a certain direction—and that direction is toward self-destruction. I don't think that's hard to show.

But the assumption that there is nothing we can do is just wrong. There is a lot that we can do. In fact, what happened in Madison illustrates that very clearly. The protesters didn't win, but it was an important demonstration. It's a basis for going further. There's plenty that we can do, but it's not going to happen by itself. If people are made to feel helpless, isolated, atomized, then power will win. These issues are pretty severe. Right now, for example, we are really facing the prospect of something like species destruction for the first time in human history.

3

UPRISINGS

CAMBRIDGE, MASSACHUSETTS (JANUARY 17, 2012)

Mohamed Bouazizi, a young street vendor in a small town in Tunisia, in despair burned himself to death.[1] *That led to what seemed to be a spontaneous uprising in Tunisia and then later in Egypt and other parts of the Arab Middle East.*

First of all, let's remember that there had been plenty going on beneath the surface. It just hadn't broken through. Take Egypt, the most important country in the region. The January 25 demonstration in Egypt was led by a fairly young, tech-savvy group called the April 6 Movement. Why April 6? The reason is that on April 6, 2008, the Egyptian labor movement, which has been quite militant and active, though suppressed, had planned to organize major strike actions at the most important industrial center in Egypt and broader solidarity actions throughout the country, but it was crushed by force by the security forces of Egyptian president Hosni Mubarak. So that's a reflection of the significant tradition of worker struggles. Though there isn't much reporting, it does seem that the Egyptian labor movement is continuing to take some pretty interesting steps, even taking over factories in some places.[2]

In the case of Tunisia, it was indeed this single act that sparked what had been long-standing active protest movements and moved them forward. But that's not so unusual. Let's look at our own history. Take the civil rights movement. There had been plenty of concern and activism about violent repression of blacks in the South, and it took a couple of students sitting in at a lunch counter to really set it off. Small acts can make a big difference when there is a background of concern, understanding, and preliminary activism.

Where do you place the rebellions that have been called the "Arab Spring" historically?

It's a triple revolt. Partly it's a revolt against Western-backed, U.S.-backed dictators throughout the region. Partly it's an economic revolt against the impact of neoliberal policies of the last several decades. And partly it's a revolt against military occupation, though most discussion of the Arab Spring leaves out two parts of the Middle East and North Africa that are under military occupation: Western Sahara and Palestine.

Actually, the so-called Arab Spring began in November 2010 in Western Sahara. Western Sahara is the last literal African colony. It's under UN jurisdiction and was supposed to be decolonized. In fact, it did move toward decolonization in 1975, but it was immediately invaded by Morocco. Morocco, mostly a French dependency, invaded and started flooding the country with Moroccans to try to overwhelm any possible independence movement. There has been a long nonviolent struggle. In November 2010, there were Arab Spring–style protests, including the creation of a tent city in one of the major cities.[3] Moroccan troops immediately came in and smashed it up. Since Morocco is a UN dependency, the Saharawi movement, a Western Sahara indigenous movement, brought a protest to the Security Council, which is responsible for decolonization. France killed it and the United States backed it.[4] So that disappeared from history.

Palestine is also under military occupation. Palestinians have made some attempts to try to join the liberation movements in the Arab world, but they were crushed pretty fast. So essentially nothing is happening in the two parts of the Middle East and North African region that are literally under Western-supported foreign occupation: France, in the case of the Western Sahara, with the United States going along, and the United States mainly in the case of Israel's occupation of Palestine.

Apart from the revolt against literal occupation, you have revolts against dictatorships and against neoliberal economics. And both of those fall into a regular pattern. So, as we discussed, Latin America has finally broken free of both political dictatorship and neoliberal policies, which had the same effect in Latin America as they had in the Middle East and North African countries—as well as here and in Europe, under slightly different modalities. They enrich a very small sector of the population while harshly punishing the rest, both in plain economic terms, such as declining real incomes, and in the quality of life, workplace freedom. You can't really impose neoliberal principles without a harsh regime. And there's been a revolt against this.

Another respect in which the revolts are similar—almost identical, in fact—is that the destructive effects of neoliberalism are very highly praised by what's sometimes called the International Monetary Fund (IMF)–World Bank–U.S. Treasury troika. In fact, in the case of Egypt, international financial elites highly praised the Mubarak dictatorship for its amazing economic performance and reforms up to just weeks before the regime crashed.

Similar things are happening in Africa, here, and in Europe. The *indignados* in southern Europe and the Occupy movements here are in a sense similar, even if they are from different worlds. The protests are not against dictatorships but against the shredding of democratic systems and the consequences of the Western version of the neoliberal system, which has had structurally consistent effects for the past thirty years: a very narrow concentration of wealth in a fraction of 1 percent of the population, stagnation for a large part of the rest, deregulation, and repeated financial crises, each one harsher than the last. The most recent financial crisis, apart from what it has done to the general population, has been absolutely devastating for the African American population. Their net worth is now one twentieth that of whites. It's the lowest it's been since statistics were first taken.[5] Average African American household net worth is down to just several thousand dollars, essentially nothing, as a result of the crash of the housing market.[6]

Talk about the role of labor in the Arab Spring.

If you look at the countries where there has been some success, Tunisia and Egypt—Tunisia more than Egypt, in fact—they both have a tradition of labor militancy. There is a close correlation between the degree of success in the Arab Spring and the participation of the labor movement. Joel Beinin, a leading scholar of Middle East and North African labor movements, pointed this out.[7] He's correct. The Tahrir Square demonstrations really became substantial and significant when the labor movement joined in. In fact, the labor movement has achieved a lot. There are now significant steps toward unification into an independent union. There had been no independent unions before. The press has been freed up. The old regime is still pretty much in place, but there's been some significant progress.

In Tunisia, one part of the population was organized: political Islam. It was repressed and crushed by the dictatorship, but it was organized. They won the parliamentary election and are introducing a moderate version of political Islam.[8] Tunisia has a major labor movement as well, which was a central part of the changes there.

In the rest of the Middle East and North African region, not much has happened. In the core countries, from the Western point of view—the oil producers, Saudi Arabia, the Emirates—there were small efforts toward joining the demonstrations, but they were very quickly crushed. In Saudi Arabia, the key country, the security presence was so overwhelming that people were afraid to go out in the streets. In Bahrain, which is not a major oil producer but is an important part of this regional system, an uprising was brutally crushed by the Saudi-led invasion, though it still goes on. Complicated things are happening in Yemen, of great concern to Saudi Arabia. Saudi Arabia appears to be supporting the former dictator. We don't have information from Saudi Arabia—it's a very closed society—but that's what it looks like.

What about Libya?

In Libya, there was an uprising. Then came, actually, two Western interventions. The first intervention was under the rubric of Security Council Resolution 1973, which called for a no-fly zone, a cease-fire, and protection of civilians. That intervention lasted maybe five minutes. The North Atlantic Treaty Organization (NATO) powers, meaning the traditional imperial powers—Britain, France, the United States—immediately joined the rebel forces and became their air force. No cease-fire, no protection to civilians. You can argue about whether they were right or wrong to do it, but the fact is they joined a rebellion to overthrow the regime. It had nothing to do with the wording of the UN resolution. Meanwhile, most of the rest of the world was trying hard to prevent a likely humanitarian catastrophe, which in fact took place, particularly at the end, when the imperial triumvirate and the rebel forces attacked the base of Libya's largest tribe, the Warfalla.[9] The attack ended up being pretty brutal and apparently has left plenty of resentment. We don't know where that will go.

At the very beginning, most of the world was calling for negotiations, diplomacy, and a cease-fire, which Gadhafi at least formally accepted. Whether it would have worked, we don't know. The African Union (AU) came out with a strong call for negotiations and diplomacy.[10] The so-called BRICS—Brazil, Russia, India, China, South

Africa—also called for that.[11] Europe was ambivalent. Germany didn't go along.[12] Turkey even tried to block the first NATO actions and later on joined in reluctantly.[13] Egypt didn't want anything to do with it.[14]

The AU is particularly interesting. Libya is an African country. The AU came out in the middle of the bombing, reiterating its call for diplomacy and making detailed proposals, in this case about a peacekeeping force.[15] They were totally dismissed, of course. You don't listen to Africans. The AU had a pretty interesting explanation of its stand. Essentially they were saying, Africa has been trying to free itself from brutal colonial rule and slavery for years. The way we've been doing it is by establishing the principle of sovereignty in order to protect ourselves from a return of Western colonization. And we have to perceive an attack on an African country over the objections of Africa, without any concern for sovereign rights, as a step toward recolonization that is very threatening to the whole continent. *Frontline* magazine in India had detailed reporting of the AU position.[16] I didn't notice a word about it here. Again, you can argue that the intervention was right or wrong, you can debate it as you like. But we might as well face what it is.

At the same time as the Obama administration cautioned the various revolutionaries in different countries to "show restraint" and said there was "no place for violence," the president praised the "unique capabilities" of the United States when it comes to enforcing a no-fly zone over Libya.[17] *Tariq Ali, in a recent article, calls Libya "another case of selective vigilantism by the West."*[18]

First of all, we should be clear that there was no Libyan no-fly zone. UN Resolution 1973 did call for a no-fly zone, but the three traditional imperial countries—Britain, France, and the United States—immediately disregarded the resolution and instantly turned to participation on the side of the rebels. So they were not imposing a no-fly zone over the rebel advances. In fact, they were encouraging them and supporting them. The United States, Britain, and France determined at once to disregard the UN resolution and to proceed to try to help the rebels overthrow the government.

Is it selective? Sure. But it's pretty predictable and very familiar. If there is a dictator who has a lot of oil and is obedient, submissive, and reliable, he's given free rein. The most important example is Saudi Arabia. There, there were supposed to be demonstrations, a "day of rage," but the government intervened with overwhelming force. Apparently not a single person was willing to appear in Riyadh. Everyone was terrified.[19]

Bahrain is particularly important in this context. It hosts the U.S. Fifth Fleet, which is by far the most powerful military force in the region, and it's right off the coast of eastern Saudi Arabia. Eastern Saudi Arabia is where most of the country's oil is. Like Bahrain, it's also largely Shiite, while the Saudi Arabian government is Sunni. By some weird accident of history and geography, the concentration of the world's energy resources is in the northern Gulf region, which is mostly Shiite and in a largely Sunni world. It's long been a nightmare for Western planners to consider the possibility that there might be some kind of tacit Shiite alliance, beyond Western control, that could take control of most of the core of the world's energy supplies.

So there was barely a tap on the wrist when Saudi Arabia led a military force into neighboring Bahrain to violently crush the protests there.[20] Backed by the Saudis, Bahraini security forces drove the protesters out of Pearl Square, where they had been camping, and even went so far as to destroy the symbol of the country, the pearl statue in the middle of Pearl Square.[21] The statue had been appropriated by the demonstrators, so army forces smashed it. They also went into a hospital and drove out a number of patients and staff.[22] That was okay. Practically no one commented on it here.

On the other hand, when you have a dictator like Mu'ammar Gadhafi, who has plenty of oil but is unreliable, it makes sense from an imperial point of view to see if you can replace him with someone more pliable and more trustworthy who will do what you want him to do. Therefore, you react differently in Libya.

In Egypt or Tunisia you follow the traditional game plan. It's as old as the hills. If there is a dictator you support who's losing control, support him until the end. If this becomes impossible, because maybe the army or the business community turns against him, shelve him, send him off somewhere, issue dramatic proclamations about your love of democracy, and then try to restore the old regime to the extent it's possible. That's exactly what's happening in Egypt. You can call it selective if you want, but it seems like rational imperialism to me—and all familiar.

In terms of these multiple uprisings throughout the Middle East, there is an embedded assumption in all the commentary that somehow the United States must control what is going on.

That's sometimes said very openly. In the *Wall Street Journal*, which tends to be franker about such things, the main political commentator, Gerald Seib, said straight out, the problem is we "haven't learned to control" these new forces.[23] The implication is that we'd better find a way to control them. That goes back sixty years to Roosevelt's planners and advisers. Adolf Berle, one of the leading liberal advisers for many presidents—

Wasn't he part of FDR's brain trust?

Yes, and then remained a major figure in the liberal political system. He said straight out, if we can control Middle East energy, that will provide us with "substantial control of the world."[24] That's no small thing.

Does the United States still have the same level of control over the energy resources of the region as it once had?

The major energy-producing countries are still firmly under the control of the Western-backed dictatorships. So, actually, the progress made by the Arab Spring is limited, but it's not insignificant. The Western-controlled dictatorial system is eroding. In fact, it's been eroding for some time. So, for example, if you go back fifty years, since then the energy resources—the main concern of U.S. planners—have been mostly nationalized. There are constantly attempts to reverse that, but they have not succeeded.

Take the U.S. invasion of Iraq, for example. To everyone except a dedicated ideologue, it was pretty obvious that we invaded Iraq not because of our love of democracy but because it's maybe the second- or third-largest source of oil in the world and is right in the middle of the major energy-producing region.[25] You're not supposed to say this. It's considered a conspiracy theory.

The United States was seriously defeated in Iraq by Iraqi nationalism—mostly by nonviolent resistance. The United States could kill the insurgents, but it couldn't deal with half a million people demonstrating in the streets. Step by step, Iraq was able to dismantle the controls put in place by the occupying forces. By November 2007, it was becoming pretty clear that it was going to be very hard to reach U.S. goals. And at that point, interestingly, those goals were explicitly stated. So in November

2007 the Bush II administration came out with an official declaration about what any future arrangement with Iraq would have to be. It had two major requirements: one, that the United States must be free to carry out combat operations from its military bases, which it will retain; and, two, "encouraging the flow of foreign investments to Iraq, especially American investments."[26] In January 2008, Bush made this clear in one of his signing statements.[27] A couple of months later, in the face of Iraqi resistance, the United States had to give that up. Control of Iraq is now disappearing before their eyes.

Iraq was an attempt to reinstitute by force something like the old system of control, but it was beaten back. In general, I think, U.S. policies remain constant, going back to the Second World War. But the capacity to implement them is declining.

Declining because of economic weakness?

Partly because the world is just becoming more diverse. It has more diverse power centers. At the end of the Second World War, the United States was absolutely at the peak of its power. It had half the world's wealth and every one of its competitors was seriously damaged or destroyed. It had a position of unimaginable security and developed plans to essentially run the world—not unrealistically at the time.

This was called "Grand Area" planning.

Yes, that was while the war was still under way. Right after the Second World War, George Kennan, head of the U.S. State Department policy planning staff, and others sketched out the details, and then they were implemented. What's happening now in the Middle East and North Africa, to an extent, and in South America substantially goes all the way back to the late 1940s. The first major successful resistance to U.S. hegemony was in 1949. That's when an event took place, which, interestingly, is called "the loss of China." It's a very interesting phrase, never challenged. There was a lot of discussion about who is responsible for the loss of China. It became a huge domestic issue. But it's a very interesting phrase. You can only lose in the relevant sense something if you own it. It was just taken for granted: we possess China—and if the Chinese move toward independence, we've lost China. Later came concerns about "the loss of Latin America," "the loss of the Middle East," "the loss of" certain countries, all based on the premise that we own the world and anything that weakens our control is a loss to us and we wonder how to recover it.

Today, if you read, say, foreign policy journals or, in a farcical form, listen to the Republican debates, they're asking, "How do we prevent further losses?" If you listen to Mitt Romney, the likely Republican presidential candidate, the way you prevent further losses is by just killing everybody who is in your way. If we don't like them, we'll kill them. In fact, that's just what he said last night.[28] That's one version. But it's the same concern: we have to maintain our control of the world.

On the other hand, the capacity to preserve control has sharply declined. By 1970, the world was already what was called tripolar economically, with a U.S.-based North American industrial center, a German-based European center, roughly comparable in size, and a Japan-based East Asian center, which was already by then the most dynamic growth region in the world. Since then, the global economic order has become much more diverse. So it's harder to carry out our policies, but the underlying principles have not changed much.

Take the Clinton doctrine. The Clinton doctrine was that the United States is entitled to resort to unilateral force to ensure "uninhibited access to key markets, energy supplies, and strategic resources."[29] That goes beyond anything that George W. Bush said. But it was quiet and it wasn't arrogant and abrasive, so it didn't cause much of an uproar. The belief in that entitlement continues right to the present. It's also part of the intellectual culture.

Right after the assassination of Osama bin Laden, amid all the cheers and applause, there were a few critical comments questioning the legality of the act. Centuries ago, there used to be something called presumption of innocence. If you apprehend a suspect, he's a suspect until proven guilty. He should be brought to trial. It's a core part of American law. You can trace it back to Magna Carta. So there were a couple of voices saying maybe we shouldn't throw out the whole basis of Anglo-American law.[30] That led to a lot of very angry and infuriated reactions, but the most interesting ones were, as usual, on the left liberal end of the spectrum. Matthew Yglesias, a well-known and highly respected left liberal commentator, wrote an article in which he ridiculed these views.[31] He said they're "amazingly naive," silly. Then he expressed the reason. He said that "one of the main functions of the international institutional order is precisely to legitimate the use of deadly military force by western powers." Of course, he didn't mean Norway. He meant the United States. So the principle on which the international system is based is that the United States is entitled to use force at will. To talk about the United States violating international law or something like that is amazingly naive, completely silly. Incidentally, I was the target of those remarks, and I'm happy to confess my guilt. I do think that Magna Carta and international law are worth paying some attention to.

I merely mention that to illustrate that in the intellectual culture, even at what's called the left liberal end of the political spectrum, the core principles haven't changed very much. But the capacity to implement them has been sharply reduced. That's why you get all this talk about American decline. Take a look at the year-end issue of *Foreign Affairs*, the main establishment journal. Its big front-page cover asks, in bold face, "Is America Over?"[32] It's a standard complaint of those who believe they should have everything. If you believe you should have everything and anything gets away from you, it's a tragedy, the world is collapsing. So is America over? A long time ago we "lost" China, we've lost Southeast Asia, we've lost South America. Maybe we'll lose the Middle East and North African countries. Is America over? It's a kind of paranoia, but it's the paranoia of the superrich and the superpowerful. If you don't have everything, it's a disaster.

The New York Times *describes the "defining policy quandary of the Arab Spring: how to square contradictory American impulses that include support for democratic change, a desire for stability and wariness of Islamists who have become a potent political force."*[33] *The* Times *identifies three U.S. goals. What do you make of them?*

Two of them are accurate. The United States is in favor of stability. But you have to remember what stability means. Stability means conformity to U.S. orders. So, for example, one of the charges against Iran, the big foreign policy threat, is that it is destabilizing Iraq and Afghanistan. How? By trying to expand its influence into neighboring countries. On the other hand, we "stabilize" countries when we invade them and destroy them.

I've occasionally quoted one of my favorite illustrations of this, which is from a well-known, very good liberal foreign policy analyst, James Chace, a former editor of *Foreign Affairs*. Writing about the overthrow of the Salvador Allende regime and the imposition of the dictatorship of Augusto Pinochet in 1973, he said that we had to "destabilize" Chile in the interests of "stability."[34] That's not perceived to be a contradiction—and it isn't. We had to destroy the parliamentary system in order to gain stability, meaning that they do what we say. So yes, we are in favor of stability in this technical sense.

Concern about political Islam is just like concern about any independent development. Anything that's independent you have to have concern about because it might undermine you. In fact, it's a little ironic, because traditionally the United States and Britain have by and large strongly supported radical Islamic fundamentalism, not political Islam, as a force to block secular nationalism, the real concern. So, for example, Saudi Arabia is the most extreme fundamentalist state in the world, a radical Islamic state. It has a missionary zeal, is spreading radical Islam to Pakistan and elsewhere, and funding terror. But it's the bastion of U.S. and British policy. They've consistently supported it against the threat of secular nationalism from Gamal Abdel Nasser's Egypt and Abd al-Karim Qasim's Iraq, among many others. But they don't like political Islam because it might become independent.

The first of the three points, our yearning for democracy, that's about on the level of Joseph Stalin talking about the Russian commitment to freedom, democracy, and liberty for the world. It's the kind of statement you laugh about when you hear it from commissars or Iranian clerics, but you nod politely and maybe even with awe when you hear it from their Western counterparts.

If you look at the record, the yearning for democracy is a bad joke. That's even recognized by leading scholars, though they don't put it this way. One of the major scholars on so-called democracy promotion is Thomas Carothers, who is pretty conservative and highly regarded—a neo-Reaganite, not a flaming liberal. He worked in Reagan's State Department and has several books reviewing the course of democracy promotion, which he takes very seriously. He says, yes, this is a deep-seated American ideal, but it has a funny history. The history is that every U.S. administration is "schizophrenic."[35] They support democracy only if it conforms to certain strategic and economic interests. He describes this as a strange pathology, as if the United States needed psychiatric treatment or something. Of course, there's another interpretation, but one that can't come to mind if you're a well-educated, properly behaved intellectual.

Within several months of the toppling of Mubarak in Egypt, he was in the dock facing criminal charges and prosecution.[36] *It's inconceivable that U.S. leaders will ever be held to account for their crimes in Iraq or beyond. Is that going to change anytime soon?*

That's basically the Yglesias principle: the very foundation of the international order is that the United States has the right to use violence at will. So how can you charge anybody?

And no one else has that right.

Of course not. Well, maybe our clients do. If Israel invades Lebanon and kills a thousand people and destroys half the country, okay, that's all right. It's interesting. Barack Obama was a senator before he was president. He didn't do much as a senator, but he did a couple of things, including one he was particularly proud of. In fact, if you looked at his website before the primaries, he highlighted the fact that, during the Israeli invasion of Lebanon in 2006, he cosponsored a Senate resolution demanding that the United States do nothing to impede Israel's military actions until they had achieved their objectives and censure Iran and Syria because they were supporting resistance to Israel's destruction of southern Lebanon, incidentally, for the fifth time in twenty-five years.[37] So they inherit the right. Other clients do, too.

But the rights really reside in Washington. That's what it means to own the world. It's like the air you breathe. You can't question it. The main founder of contemporary IR theory, Hans Morgenthau, was really quite a decent person, one of the very few political scientists and international affairs specialists to criticize the Vietnam War on moral, not tactical, grounds. Very rare. He wrote a book called *The Purpose of American Politics*.[38] You already know what's coming. Other countries don't have purposes. The purpose of America, on the other hand, is "transcendent": to bring freedom and justice to the rest of the world.[39] But he's a good scholar, like Carothers. So he went through the record. He said, when you study the record, it looks as if the United States hasn't lived up to its transcendent purpose. But then he says, to criticize our transcendent purpose "is to fall into the error of atheism, which denies the validity of religion on similar grounds"—which is a good comparison.[40] It's a deeply entrenched religious belief. It's so deep that it's going to be hard to disentangle it. And if anyone questions that, it leads to near hysteria and often to charges of anti-Americanism or "hating America"—interesting concepts that don't exist in democratic societies, only in totalitarian societies and here, where they're just taken for granted.

4

DOMESTIC DISTURBANCES

CAMBRIDGE, MASSACHUSETTS (JANUARY 17, 2012)

As someone who is interested in the political deployment of language, you must appreciate the irony of "occupy" and "occupation," which are extremely negative terms, being used in a very positive way by the Occupy movement.

It's an interesting usage, and it took off. Occupy now means taking something over for popular goals. Occupying public space has been a very successful tactic. I would have never guessed it would have worked, frankly. There is an incipient movement called Occupy the Dream. It was formed by representatives of the Occupy movements and surviving leaders of the original civil rights movement, including Benjamin Chavis.[1] The dream that they're talking about is not the one that people refer to on

Martin Luther King Day, the civil rights dream. It's King's real dream: end war, end poverty, deal with the real suffering of people, not just civil rights, which is hard enough.

There has been an increase in the use of terms such as "income inequality," "concentrations of wealth," "CEO pay," "poverty," "unemployment" since the Occupy Wall Street movement began in September 2011.

The idea of the 1 percent and 99 percent has become common. The Occupy movement has succeeded in tapping feelings, attitudes, and understandings that have been latent, hidden right below the surface. They brought it out. All of a sudden it exploded. It's interesting, if you take a look at the business press, the *Financial Times*, which is the most important business daily in the world, has been surprisingly sympathetic to the Occupy movements. Not to their longer-term goals—they don't talk about those—but the short-term ones. They use a lot of this imagery now quite freely, and in quite a sympathetic way.

There are enormous propaganda efforts to try to denigrate it and undermine the movement, to say it's the politics of envy. Why don't you shower and get a job? And this has its effect, undoubtedly. But still Occupy lit a spark, and it's changed the substance, as well as the tone, of national discourse on crucial issues.

But as with any movement, you have to keep thinking through what you're doing. The Occupy tactic has been extremely successful. It was a brilliant tactic, not just for raising issues but also for creating communities—something very important in a society like ours, which is so atomized. People are alone. They sit alone in front of their TV set. You don't "consult your neighbor," to use the old Wobbly phrase. That atomization is a technique of control and marginalization. One of the real achievements of Occupy has been to bring people together to form functioning, supportive, free, democratic communities—everything from kitchens to libraries to health centers to free general assemblies, where people talk freely and debate. It's created bonds and associations that, if they last and if they expand, could make a big difference.

But a tactic has a half-life. It works for a while, and then you see diminishing returns. It's inevitable. So it's important at some stage, maybe now, to ask whether the Occupy tactic has essentially lived its life and it's time to turn to something else, like the Occupy the Dream movement. Around New York, Boston, other places, there have been Occupy the Hood movements in poor and minority neighborhoods, where people get together to deal with their own problems, drawing inspiration from the downtown Occupy movements but saying, "We'll do it here." That's really important.

Also, I think there should be a lesson from Tunisia and Egypt, and from the 1930s here. Unless the labor movement is revitalized and becomes a core part of the movement, I don't think it's going to get very far. Revitalizing the labor movement may seem like a real long shot, if you take a look at the country today, but conditions now are actually no worse than they were in the 1930s. Remember that by the 1920s the American labor movement, which had been militant and successful, had been virtually crushed.

The Red Scare and the Palmer raids had crushed labor and independent thought and created an end-of-history mentality, a utopia of the masters, rather like the early 1990s. But the labor movement was resurrected. In fact, if you go back to the 1920s, visitors from abroad, including conservatives, were just appalled at the treatment and the status of American workers. There was nothing like it in other major industrial countries. But in the 1930s labor revived and you had the formation of the CIO, sit-down strikes. It could happen again. The seeds of it are there.

In 1968, a slogan was raised in France to "demand the impossible." What do you remember about that particular period that might have some relevance to what's going on today?

What happened in France was significant. The most significant part, at least for me, was the fact that there was an incipient student-worker alliance, which could have meant something. Actually, it turned out it didn't, but it really could have meant something. That's an example of a spark that didn't lead to a conflagration.

In order to mount resistance and challenge power, it's necessary to overcome the barrier of fear. It seems that the Occupy movement has done that.

It has. It's costly to oppose power. No matter if you're a graduate student, a child in school questioning something that's happening, a union organizer, or a political dissident, whatever you may be, it's going to carry a personal cost. Power systems, whatever they are, very rarely abdicate their power cheerfully. They usually resist. In a society like ours, they have many means at their disposal. We have a very class-conscious business class in the United States. They're always fighting a bitter one-sided class war and if they meet any opposition they will react. So yes, there's a cost. And fear is understandable. If you attempt to organize a union at some workplace, you can be easily subjected to punishment. The punishments are illegal, but when you have a criminal state, that doesn't matter. The state doesn't enforce the laws. In fact, just the very act of breaking out of discipline to begin to organize people carries a cost.

So fear is understandable. Nowadays it's being enhanced by pretty severe attacks on basic civil liberties. A system of control and repression is in place—it's not being excessively used, but it's in place and it can be quite punitive.

Indefinite military detention, for example.

The new National Defense Authorization Act[2] isn't as bad as it's been described on the Internet by some, but it's bad. Essentially, it codifies practices that have been carried out regularly by the Bush II and Obama administrations without particular objection. In fact, they're bipartisan. But now these practices have been codified, and that's worse. Also, the act allows for the military to be involved in domestic policing, which violates principles that go back to the late nineteenth century. And it makes military detention mandatory for people who are called terrorists or enemy combatants. For U.S. citizens, military detention is left in the law as an option but is not mandatory.[3] All of those are dangerous steps.

Still, I don't think the act is the worst attack on civil liberties under the Obama administration. There are worse ones. Maybe the worst is the Supreme Court decision in *Holder v. Humanitarian Law Project*.[4] The case, which didn't get too much attention, was brought to the court by the Obama administration and argued by former solicitor general Elena Kagan, his latest Supreme Court appointment. The Humanitarian Law Project was giving advice to a lot of groups, including some that are on the official U.S. State Department list of Foreign Terrorist Organizations.[5] They were talking to them about strategies of nonviolence.[6] The Obama administration argued in the Supreme Court that advice is "material support," and won. There already were laws against material assistance to groups on the terrorist list. You can't give them arms. But Obama expanded it to talking. So, for example, the wording of the judgment suggests that if you talk to somebody they call a terrorist and urge them to turn to nonviolence, you're guilty of giving material assistance to terrorist groups. The potential scope of that is incredible. These are executive decisions—without review, without recourse.

If you look at the record of who is designated a terrorist, it's shocking. Maybe the most extreme case is Nelson Mandela, who just got off the terrorist list about four years ago.[7] The Reagan administration, which supported the apartheid regime in South Africa right to the end, condemned the African National Congress as one of "the more notorious terrorist groups" in the world.[8] So Mandela is a terrorist because they say so. He's only now for the first time free to come to the United States without special authorization.[9] Saddam Hussein was taken off the terrorist list in 1982 so the United States could provide him with agricultural and other support that he needed.[10] The whole record is grotesque.

But extending the concept of material support to conversation—most of us could be tried under that. And the ruling was applied right away. As soon as the Supreme Court case was decided, the FBI was sent to raid apartments in Chicago and Minneapolis to collect information about people who were suspected of giving material support for the Palestinian groups and Revolutionary Armed Forces of Colombia (FARC), like maybe urging them to negotiate and turn to nonviolence.[11] That's a pretty severe attack on civil liberties.

So there are reasons for fear. The government has instruments at hand, which it shouldn't have.

We're soon going to be commemorating the eighth century of Magna Carta. Magna Carta was a huge step forward. It established the right of any freeman—later extended to every person—to be free from arbitrary persecution. It established the presumption of innocence, the right to be free from state persecution, and the right to a free and fair speedy trial. That later was expanded into the doctrine of habeas corpus and became part of the U.S. Constitution. This is the foundation of Anglo-American law and one of its highest achievements, but it's now being cast to the winds.

One of the most remarkable examples is of Omar Khadr, the first Guantánamo case to come to a military commission—not a court—under Obama. The charge was that he had tried to resist an attack on his village by American soldiers when he was a fifteen-year-old boy.[12] That's the crime. A fifteen-year-old tries to defend his village from an invading army. So he's a terrorist. Khadr had been kept in Guantánamo and, before that, Bagram in Afghanistan for eight years. I don't have to tell you what Guantánamo is like. He finally came to a military commission, where he was given a choice: either plead not guilty and stay here forever or plead guilty and just spend another eight years in detention.[13] This violates every international convention that you can think of, including laws on treatment of juveniles. Of course, it grossly violates any principle. He was fifteen. But there was no public outcry.

In fact, particularly striking in some ways is that Khadr is a Canadian citizen. Canada could extradite him and free him if it wanted to, but they didn't want to step on the master's toes.[14]

Talk about the dangers of sectarianism, which historically drove a number of wedges into social movements in the 1960s. Some of this sectarianism was engineered by the state through COINTELPRO, the Federal Bureau of Investigation's Counter Intelligence Program, and other efforts.

Sectarianism is very serious. The core of U.S. popular activism in the 1960s was the civil rights movement. But by the mid-1960s it had basically shattered. The Student Nonviolent Coordinating Committee and Students for a Democratic Society were the center of a lot of the student activism and other activism of young people. Around 1968, the student Left split into two major groups. One of them was Progressive Labor (PL), which was Maoist. So PL says, "Let's stand outside the G.E. factory in Lynn, Massachusetts, and hand out leaflets to recruit them for a Maoist revolution." I'm being a little unfair, but that's basically what it was. The other split was the Weathermen, which said, "Things are so awful and horrible that we have to start a revolution. And the way we do it is by breaking windows of banks and attacking people"—robbing Brinks trucks, things like that. It's hard to know which was more destructive. They were both destructive.

It was a real struggle to help young people escape these tendencies. Some did on their own, but a lot were caught up in them. There were a number of personal tragedies. Friends of mine spent years in jail as a result. And it effectively destroyed the movement.

COINTELPRO was part of the story, but we shouldn't exaggerate. A lot of the sectarianism was coming from inside.

One of the criticisms leveled against the Occupy Wall Street movement is that it is leaderless, anarchic, nonideological. What do you think about its decision-making process, which is nonhierarchical. These general assemblies, for example, operate on consensus.

Consensus certainly has its value, but it's also dangerous. All of us who have been around for years know that consensus decisions can turn out to be highly authoritarian. Some small group will be really dedicated to taking the movement over, and they'll hang around after everybody is bored silly and end up running it. That happens over and over. So consensus can be a good thing, but you've got to understand its limits.

Without any leadership, more or less spontaneously the movement has developed a "let a hundred flowers bloom" mentality, which I think is a good thing. They didn't develop a party line, like, say, the old Communist Party. Or to take a contemporary analogue, the Republican Party. The Republican Party today has a catechism. If you want to be a candidate, with very rare exceptions, you have to repeat the catechism in lockstep uniformity: global warming isn't happening, no taxes on the rich.

There are about ten things that you have to repeat, whether you believe them or not. Anybody who departs from them is in trouble. Part of the catechism is, if somebody is out there that we don't like or we think might harm us, "we kill them," as Romney's put it.[15] One person in the Republican debate, Ron Paul, said, "Maybe we ought to consider a golden rule . . . in foreign policy," treat others the way we want them to treat us.[16] He was practically booed off the stage. That's reminiscent of the old Communist Party.

The Occupy movements are quite right to try to avoid this quasi-totalitarian structure. On the other hand, consensus can go too far, like any other tactic. I think the criticism that Occupy hasn't come up with actual proposals or demands is just not true. There are lots of proposals that have come out of Occupy. Many of them are quite feasible, within reach. In fact, some even have mainstream support from places like the *Financial Times*, things like a financial transaction tax, which makes good sense.

That's the former Tobin tax, put forward by the economist and Nobel laureate James Tobin, sometimes called the Robin Hood tax.

Yes. A financial transaction tax would make a big difference in some countries, if it were done properly. The absolute refusal to tax the superrich is another part of the Republican catechism. Going after that—and dealing with radical inequality—makes perfectly good sense.

So does creating jobs. The basic problem we face is not a deficit but rather joblessness. A majority of the population agrees with that.[17] But the banks don't agree, so therefore it's not discussed in Washington.

We could have a reasonable health care system, like other industrial countries. Not exactly utopian. Again, fighting for that makes perfectly good sense. A single-payer health care system has a lot of popular support, but the financial institutions are against it, so it's not even discussed. A national health care system would, incidentally, eliminate the deficit, among other things—not that the deficit is all that important.

There are further goals I don't think are unfeasible but could be revolutionary in import. So, for example, if a multinational corporation is shutting down an efficient manufacturing installation because it doesn't make enough profit for them and they would rather shift it to China, the workforce and community could decide that they want to take it over, purchase it, direct it, and keep it running. In fact, that's something proposed in standard works of business economics, which point out that there is no law of economics or capitalism that says firms have to act in the interest of shareholders, not stakeholders. The stakeholder is anybody their actions have an impact on: the workforce, the community, others.

The Occupy movement could at least be as imaginative as a standard business economics text. If they pursue that, it could lead to quite far-reaching changes.

The sociologist Immanuel Wallerstein says, "Capitalism is at the end of its rope."[18] *Is it too soon to be talking about the end of capitalism?*

I don't even know what it means. First of all, we've never had capitalism, so it can't end. We have some variety of state capitalism. If you fly on an airplane, you're basically flying in a modified bomber. If you buy drugs, the basic research was done under public funding and support. The high-tech system is permeated with internal controls, government subsidies. And if you look at what are supposed to be the growing alternatives, China is another form of state capitalism. So I don't know what's supposed to be ending.

The question is whether these systems, whatever they

are, can be adapted to current problems and circumstances. For example, there's no justification, economic or other, for the enormous and growing role of financial institutions since the 1970s. Even some of the most respected economists point out that they're just a drag on the economy. Martin Wolf of the *Financial Times* says straight out that the financial institutions shouldn't be allowed to have anything like the power they do.[19] There's plenty of leeway for modification and change. Worker-owned industries can take over. There's interesting work on this topic by Gar Alperovitz, who has been right at the center of a lot of the organizing around worker control.[20] It's not a revolution, but it's the germ of another type of capitalism, capitalism in the sense that markets and profit are involved.

Howard Zinn once commented, "There is a basic weakness in governments—however massive their armies, however wealthy their treasuries, however they control the information given to the public—because their power depends on the obedience of citizens, of soldiers, of civil servants, of journalists and writers and teachers and artists. When these people begin to suspect they have been deceived, and when they withdraw their support, the government loses its legitimacy, and its power."[21] *He also wrote that people "know with supreme clarity—when their attention is not concentrated by the government and the media on waging war—that the world is run by the rich."*[22]

That's basically correct. And incidentally, without taking anything away from Howard, it's an old principle. I think maybe the classic formulation was by David Hume in "Of the First Principles of Government," where he pointed out that "Force is always on the side of the governed."[23] Whether it's a military society, a partially free society, or what we—not he—would call a totalitarian state, it's the governed who have the power. And the rulers have to find ways to keep them from using their power. Force has its limits, so they have to use persuasion. They have to somehow find ways to convince people to accept authority. If they aren't able to do that, the whole thing is going to collapse.

When coercion doesn't work anymore, you have to turn to persuasion. In the rich, developed societies this has become an art form. In Britain and the United States, the freest societies about century ago, it was very clearly recognized by the leadership, the Tory Party in Britain, intellectuals in the United States, that the limits of coercion had been reached. People had won too much freedom—parliamentary labor parties, labor unions, women's rights groups. So you had to turn to control of attitudes and opinion. That's the origins of the public relations industry. Edward Bernays, the guru of the U.S. public relations industry, a liberal progressive, expressed the standard view, which wasn't novel to him: "Ours must be a leadership democracy administered by the intelligent minority who know how to regiment and guide the masses."[24] We somehow have to persuade or change the attitudes of the population so they will be willing to hand power over to us. Whoever presents these views is always part of the "intelligent minority." And the way we do it is through propaganda. The term was used openly then. In fact, Bernays titled his book *Propaganda*. The word took on bad connotations in the 1930s, but before that it was used freely. Now it's called advertising or public relations.

Those are the foundations of the industries of control of opinions and attitudes, driving people to consumerism and marginalizing them in various ways. Huge resources are devoted to this. Marketing is mostly a form of propaganda. If anybody believed in markets, which only ideologues do, but if, say, business believed in markets, they wouldn't do anything like the marketing they do today. If you take an economics course, they teach you that markets are based on informed consumers making rational choices. But business devotes huge resources to trying to create uninformed consumers who make irrational choices. It's obvious as soon as you look at an advertisement. If you had a market system, General Motors, let's say, would put up a thirty-second ad on television saying, "Here are the characteristics of the cars we're selling next year." They obviously don't do that, because they want to undermine markets.

In fact, political and business leaders want to undermine democracy the same way. Democracy, you learn in eighth grade, is made up of informed voters making rational choices, but the political parties certainly don't believe in that. That's why they have slogans, rhetoric, public relations displays, extravaganzas, anything but saying, "Here's what I'm going to do. Vote for me." So the fear and hatred of markets and of democracy basically have the same roots.

Again, the point is correct. Hume is the first person who articulated it clearly, as far as I know. The public does have power, and it's the task of the powerful and their minions—the priests, the intellectuals, others—to try to marginalize them, to get the public away from power. We have Walter Lippmann, the famous leading public intellectual of the twentieth century, also a progressive, saying that we've got to protect the responsible men, the intelligent minority, from the "trampling and roar of the bewildered herd."[25] That's what the huge public relations industry is devoted to.

In late 2011, New York Times *columnist David Brooks reported that a Gallup poll showed that in answer to the question "Which of the following will be the biggest threat to the country in the future—big business, big labor, or big government?" close to 65 percent of respondents said the government and 26 percent said corporations.*[26] *Is that an example of the persuasion and manufacturing of consent that you alluded to?*

If you look a little bit beyond that question and you ask, "What do you want the government to do?" the answer will be, "Stop bailing out the banks. That's why I hate the government. Don't bail out the banks. Stop freeing the rich from taxes. I want more taxes on the rich. Increase spending on health and education." And so on down the line. So yes, the question is framed so that people like David Brooks can draw this conclusion.

Take welfare. There's strong public opposition to welfare. On the other hand, there's strong public support for what welfare does. So if you ask the question, "Should we spend more on welfare?" No. "Should we spend more on aid to women with dependent children?" Yes.[27] That's successful propaganda. Welfare has been successfully demonized. Reagan took a big step forward on that, sort of constructing an image of welfare as meaning a rich, black woman who drives to the welfare office in her chauffeured limousine and takes away your hard-earned money. Nobody is in favor of that, so no welfare. But what about a mother with a child that she can't feed? Oh, yes, we're in favor of helping her.

In fact, if you look at the 1960s, there were significant changes in the way these issues were conceived. A useful study of this shift just came out in *Political Science Quarterly*.[28] The New Deal conception was that support for people's needs was a right. So, say, a mother with dependent children had a right to food for her children. That began to shift in the 1960s. As the welfare system was expanded, a shift began toward the conception that you can get support but you really ought to be working, ultimately leading to the move from welfare to workfare. By the time you get to Clinton, the right to food for your children is not really a right.[29] It's only something until you get a job, which is what you ought to be doing. This is based on the idea that taking care of children isn't work. It's an amazing conception. Anyone who has taken care of children knows it's work, hard work. Even from an economic point of view, adopting the rather ugly terminology of standard economics, it creates what's called "human capital." In economics courses, human capital, the quality of the workforce, is terribly important. How do you get human capital in a four-year-old child? When the mother is at home taking care of him, not letting him run out in the streets while she's washing dishes in a restaurant. And, of course, there's almost no support for the working family, so you destroy the family. It's a very striking shift in mentality.

The driving force behind these changes is people who claim that they are fighting for "family values." The people who call themselves conservatives say, "We have to maintain family values by preventing women from having a choice as to whether they will have children, and then by not giving them any support when they have to take care of their children. That's how we preserve family values." The internal contradictions are amazing.

Talking about the mechanisms of domestic control reminds me of Aristotle's comments about democracy. What did he have to say about democracy?

In his book *Politics,* which is the foundation of the study of political systems, and very interesting, Aristotle talked mainly about Athens. But he studied various political systems—oligarchy, monarchy—and didn't like any of them particularly. He said democracy is probably the best system, but it has problems, and he was concerned with the problems. One problem that he was concerned with is quite striking because it runs right up to the present. He pointed out that in a democracy, if the people—people didn't mean people, it meant freemen, not slaves, not women—had the right to vote, the poor would be the majority, and they would use their voting power to take away property from the rich, which wouldn't be fair, so we have to prevent this.[30]

James Madison made the same point, but his model was England. He said if freemen had democracy, then the poor farmers would insist on taking property from the rich.[31] They would carry out what we these days call land reform. And that's unacceptable. Aristotle and Madison faced the same problem but made the opposite decisions. Aristotle concluded that we should reduce inequality so the poor wouldn't take property from the rich. And he actually proposed a vision for a city that would put in place what we today call welfare-state programs, common meals, other support systems. That would reduce inequality, and with it the problem of the poor taking property from the rich. Madison's decision was the opposite. We should reduce democracy so the poor won't be able to get together to do this.

If you look at the design of the U.S. constitutional system, it followed Madison's approach. The Madisonian system placed power in the hands of the Senate. The executive in those days was more or less an administrator, not like today. The Senate consisted of "the wealth of the nation," those who had sympathy for property owners and their rights. That's where power should be. The Senate, remember, wasn't elected. It was picked by legislatures, who were themselves very much subject to control by the rich and the powerful. The House, which was closer to the population, had much less power. And there were all sorts of devices to keep people from participating too much—voting restrictions and property restrictions. The idea was to prevent the threat of democracy. This goal continues right to the present. It has taken different forms, but the aim remains the same.

5

UNCONVENTIONAL WISDOM

CAMBRIDGE, MASSACHUSETTS (JANUARY 20, 2012)

Talk about the economic crisis in Europe and its impact on the United States. The euro zone, with seventeen countries, has a unitary monetary system, but the European Union itself has twenty-seven member countries.

It's pretty hard to explain what the European Central Bank (ECB) is doing except in terms of conscious class war. A pretty broad spectrum of economists, including those who are pretty conservative, recognize that the worst possible policy during a recession is austerity. You have to stimulate economies during a recession, not cause them to decline. But the European Central Bank is rigidly adhering to austerity programs, under mostly German influence. The U.S. Federal Reserve, at least in principle, has a dual mandate: one of them is to control inflation, the other is to maintain employment. They don't really do it, but that's the mandate. The European Central Bank has only one objective, to control inflation. It's a bankers' bank, nothing to do with the population. They have an inflation target of 2 percent, and you're not allowed to threaten that.[1] In fact, there is no threat of inflation in Europe. But they insist on not carrying out any stimulus or anything like quantitative easing or other measures that might increase growth.

The effect is that the weaker countries in the European Union are never going to be able to get out of their debt under these policies. In fact, debt levels are getting worse. As you cut down growth, you cut down the possibility of debt repayment. Hence they sink deeper into misery. Under ECB policies, Greece and Spain, in particular, are being punished and driven down.

It's hard to think of a reason for this other than class war. The effect of the policies is to weaken welfare-state measures and to reduce the power of labor. That's class war. It's fine for the banks, for financial institutions, but terrible for the population.

How will this effect the United States, which is a major trading partner with Europe?

Not only is it a major trading partner, but U.S. banks are heavily invested in European institutions.[2] So yes, they may suffer from it, too. In fact, what's been happening is that there's been a flow of investor funds to the United States, to Treasury securities, which are regarded as a safe haven now, which has a mixed effect for the United States.[3] It tends over time to raise the value of the dollar and harm exports. So it's not good for a healthy economy. But, as usual, there are winners and losers. So far the banks are making out okay.

Economist Richard Wolff has been traveling around Europe. He said in an interview I did with him in New York that this German-driven economic policy "is accomplishing for Germany what Hitler tried and failed to achieve—a Europe whose dominant center is in Berlin."[4]

There's something to that. Ever since the economic recovery began in the postwar period, the European economy has been basically German-based. Germany has the strongest economy in the region. It remains a major manufacturing center and even an export center. It's the powerhouse of Europe. And all these policies just make it more powerful. On the other hand, they may be killing the goose that lays the golden egg, because they've relied pretty heavily on the export market of the euro zone. If that collapses, German industry will take a hit. But Wolff is basically right. As I said, in Greece it's particularly striking, because they fought really hard to try to free themselves from Hitler's domination.

Let's talk about Turkey, which has been trying for years, without success so far, to get into the European Union. A front-page New York Times *article notes, "Charges Against Journalists Dim the Democratic Glow in Turkey." Turkish human rights advocates say the crackdown on journalists "is part of an ominous trend. . . . The arrests threaten to darken the image of [Prime Minister Recep Tayyip Erdoğan], who is lionized in the Middle East as a powerful regional leader who can stand up to Israel and the West." According to this report, "There are now 97 members of the news media in jail in Turkey, including journalists, publishers and distributors, according to the Turkish Journalists' Union, a figure that rights groups say exceeds the number detained in China."*[5] *One of those imprisoned is Nedim Şener, an award-winning journalist for his reporting on the murder of Hrant Dink, a prominent Turkish Armenian journalist who was assassinated in Istanbul in January 2007.*

Since you brought up irony, I should note, first of all, that this report in the *New York Times* has ample ironic connotations. What's going on in Turkey now is pretty bad. On the other hand, it doesn't begin to compare with what was going on in the 1990s. Then the Turkish state was carrying out a major terrorist war against the Kurdish population: tens of thousands of people killed, thousands of towns and villages destroyed, probably millions of refugees, torture, every atrocity you can think of.[6] The *Times* barely reported it. It certainly didn't report the fact that 80 percent of the weapons were coming from the United States and that Clinton was so supportive of the atrocities that in 1997, when they were peaking, he sent more arms to Turkey that single year than in the entire Cold War period combined.[7] That's pretty serious, but you won't find it in the *New York Times*. Its correspondent in Turkey, Stephen Kinzer, barely reported anything. Not that he didn't know. Everybody knew.

So if the *Times* is upset about human rights violations, we can take the reaction with a grain of salt. Now reporters are willing to highlight the human rights violations because Turkey has been standing up to the United States. And that they don't like. Erdoğan's popularity in the Middle East does not make him popular in the United States. He's by far the most popular figure in the Arab world, whereas Obama's popularity is actually lower than Bush II's, which is quite a trick.[8]

Turkey has taken a fairly independent role in world affairs, which the United States doesn't like at all. The country is increasing trade relations with Iran.[9] Turkey and Brazil carried out a major crime. They succeeded in getting Iran to agree to a program of transferring the low-enriched uranium out of Iran, which happened to virtually duplicate Obama's proposal.[10] In fact, Obama had actually written a letter to Luiz Inácio Lula da Silva, the Brazilian president, urging him to proceed with such a plan, mainly because Washington assumed that Iran would never agree, and then they could use this refusal as a diplomatic weapon and gain more international support for sanctions.[11] But Iran did agree. There was great anger here, because any agreement might undermine the push for sanctions, which is what Obama was really after.

And there are other sources of U.S. hostility to Turkey. For example, Turkey, which is a NATO power, interfered with NATO's early efforts to carry out the bombing of Libya.[12] Washington didn't like that either.

So now it's appropriate to condemn human rights violations in Turkey. And they do exist. Actually, there was quite considerable progress in Turkey over human rights over the past ten years, but the last couple of years have been pretty unpleasant. There's been regression. Cynicism aside, it's correct to protest the abuses in Turkey.

In March 2011, Orhan Pamuk, a leading Turkish writer and Nobel Prize winner, was fined for his statement in a Swiss newspaper that "we have killed 30,000 Kurds and 1 million Armenians."[13]

I was in Turkey a year ago at a conference on freedom of speech. A large part of it was devoted to talks by Turkish journalists, describing their activities in trying to expose the Hrant Dink murder, the atrocity against the Armenians, the repression of the Kurds. These are very courageous people. It's not like a *New York Times* correspondent, who could write about these topics if he wanted to and would suffer no consequences. Maybe he would be censured by the editors, but these guys can be sent to jail and undergo torture. That's serious. But they talk openly and strikingly.

In fact, one of the most interesting things about Turkey—again, ironically—is that the European Union says that Turkey can't join because it doesn't meet our high standards of human rights.[14] Turkey is about the only country I know of in which leading intellectuals, journalists, academics, writers, professors, and publishers not only constantly protest the atrocities of the state but regularly carry out civil disobedience against it. I actually participated to an extent when I went there ten years ago. There's nothing like that in the West. They put their Western counterparts to shame. So if there are lessons to be learned, I think it's in the other direction. Frankly, I've never thought that Turkey would be admitted into the EU, mainly on racist grounds. I don't think Western Europeans like the idea of Turks walking around freely on their streets.

How do Turkish-Israeli relations influence Washington, with the 2010 Israeli commando raid in international waters on a Turkish ship killing nine Turks, one of whom was an American citizen?[15]

Turkey was the only major country, certainly the only NATO country, to have protested very sharply against the U.S.-Israeli attack on Gaza in 2008–09.[16] And it was a U.S.-Israeli attack. Israel dropped the bombs, but the United States backed it, with Obama's approval.[17] Turkey came out very strongly in condemnation. In a famous incident in Davos at the World Economic Forum, Erdoğan spoke out strongly against the attack while Shimon Peres, the Israeli president, was onstage with him.[18] Of course, the United States didn't like that. Having cordial relations with Iran and condemning Israeli crimes does not make you a favored figure at Georgetown cocktail parties.

And now there's a report that Israel, which has long denied the Armenian genocide, is considering a resolution condemning it, primarily to irritate the Turks, who they know are hypersensitive to any mention of the Armenian genocide.[19]

It cuts both ways. Israel and Turkey were pretty close allies. In fact, Turkey was Israel's closest ally, apart from the United States. Their connection was kept under cover, but it was perfectly clear from the late 1950s on. It was very important for Israel to have a powerful non-Arab state ally. Turkey and Iran under the shah were very close to Israel. At that time, Israel refused to allow any discussion of the Armenian genocide.[20]

In 1982, Israel Charny, somebody I knew as a kid in Hebrew camp, organized a Holocaust conference in Israel.[21] He wanted to invite someone to talk about the

Armenian atrocities, and the government tried to block it. In fact, they pressured Elie Wiesel, who was supposed to be the honorary chair, to resign.[22] The conference organizers went ahead with it anyway, over strong government opposition.

At that time, Turkey was an ally, so you didn't talk about the Armenian genocide. Now, as you say, relations are frayed, so you can sort of stick it to the Turks. You can talk about it now. In fact, Israel's behavior has been pretty remarkable. One of the incidents that didn't get much publicity here but really bothered the Turks was a meeting between the Turkish ambassador to Israel, Ahmet Oğuz Çelikkol, and Danny Ayalon, the deputy foreign minister. He called in the Turkish ambassador and they set up a photo op with him sitting on a very low chair and Ayalon sitting on a higher chair above him.[23] And then the photographs are publicized all over. Countries don't act like that. It's very humiliating.

That's only one of a series of events which actually, from Israel's own strategic point of view, are not very brilliant. The Turkish-Israeli military strategic, trade, and commercial relationship has been pretty significant. Again, we don't really know the details, but for years Israel has cooperated with Turkey on military training and used its airspace for preparations for possible aggression in the Middle East.[24] If they sacrifice that, it's serious.

The Kurds are possibly the largest single ethnic group in the world without a nation-state. They have gained some semi-autonomy in northern Iraq? How viable is that?

It's fragile. There's a lot of repression and corruption in northern Iraq. Furthermore, their economy is not really viable. They're landlocked. If they don't have significant support from the outside, they can't be sustained for long. They're also surrounded by enemies, Iran on one side, Turkey on the other, and Arab Iraq as well. There's a connection to Syria, but that doesn't help much. So the Kurdish region in the north of Iraq exists by the tolerance of the great powers, primarily the United States, which could be withdrawn.

The United States has repeatedly sold the Kurds out over the years.[25] It sold them out to Saddam Hussein in the 1970s and again in the 1980s. During Saddam Hussein's atrocities against the Kurds, the U.S. government tried to silence them. The Reagan administration tried to blame the atrocities on Iran. The Kurds have an old saying, which goes something like, "Our only friends are the mountains," meaning we can't rely on outsiders for support. If you look at their history, they have plenty of reason to believe that.

One of the few American journalists to have really worked in the area, Kevin McKiernan, once described a mountain in northern Iraq called Mount Qandil. He said it has two sides: on one side there are terrorists, on the other side there are freedom fighters.[26] They're exactly the same people: they're Kurdish nationalists. But one side faces Turkey, so they're terrorists. The other side faces Iran, so they're freedom fighters.

I was just about to ask you about Iran. The bellicose talk about Iran seems to ebb and flow like the tide. Every few months we hear new reports of a potential U.S. or Israeli attack on Iran.

The rhetoric stays high. On the other hand, as far as evidence is available, the U.S. and Israeli intelligence and top military commands are not eager to be involved in a military campaign against Iran.[27] However, when you ratchet up the tension, something can happen, if only by accident. It's happened many times in the past. You can easily think of scenarios. There might be a confrontation between a small Iranian vessel carrying missiles and an American aircraft carrier. Who knows where that would lead?

And Iran is likely to retaliate pretty soon for the war now taking place against it. Because there already is a war on Iran. When you assassinate scientists and tighten sanctions to the point where they're purposely and openly strangling the economy, that's aggression.[28] It amounts to a blockade. In fact, high U.S. military officials consider those measures aggression if they're conducted against the United States. There was an analysis a couple of years ago that came out from a group of top international military thinkers, including two retired NATO generals, discussing strategic issues, defining threats to the United States that we would regard as aggression. One of them was the use of financial institutions to harm the U.S. economy.[29] That's aggression. We can respond with force. They also added that we shouldn't refrain from the first use of nuclear weapons.[30] Generalize those principles and Iran might react. If the Iranian leadership concludes that they have nothing more to lose—their economy is being strangled, their political control is going to be destroyed—they might go for broke.

Information like the U.S. support for Saddam Hussein throughout the Iran-Iraq War has gone down Orwell's memory hole.[31] *It reminds me of a comment you made about historical amnesia. You said, "Historical amnesia is a dangerous phenomenon, not only because it undermines moral and intellectual integrity, but also because it lays the groundwork for crimes that still lie ahead."*[32]

If you don't recognize your own crimes, there's no impediment to continuing them. There's a pretty dramatic example of that right at this moment. This happens to be the fiftieth anniversary of John F. Kennedy's decision to launch the war against South Vietnam. Forgetting the fiftieth anniversary of the launching of one of the major atrocities in post–Second World War history is pretty severe. But almost nobody has noticed it. I don't think we'll hear a word about it. And, yes, that opens the way to further aggression.

One topic that often comes up in the media and among policy makers is the instability of Pakistan and the vulnerability of its nuclear arsenal.

Here in the United States the discussion is uniformly about how Pakistan can't be trusted and is not a reliable ally. Suppose that the Russians had said in the 1980s, "Pakistan is not a reliable ally. We have to do something about it." That's when Pakistan was the center of the U.S. support for arming and training of the mujahideen, the guerrillas who were fighting the Russians in Afghanistan. The major experts on Pakistan, including mainstream military historians and South Asian experts, say the attitude of Pakistanis toward the Taliban today is pretty similar to their attitude toward the mujahideen in the 1980s.[33] They don't like them, they want them to stay out of their hair, but they regard them as fighting a war against a foreign invader. So there's apparently overwhelming opposition in Pakistan to the pressure to take part in an American war against people who they regard as defending their country.[34]

The United States is constantly carrying out military attacks in Pakistan. There was another one yesterday. A drone killed an alleged al-Qaeda leader who was supposedly planning actions against the United States.[35] Who knows? But the Pakistanis certainly don't like it. They don't like being bombed, no matter where it is, even if it's in the tribal areas. They're very bitter about the invasion and assassination of bin Laden, rightly so. And, in fact, the Pashtun population, which crosses the border of Afghanistan and Pakistan, has never accepted the Durand Line, the British-imposed boundary that cuts right through their territory.

Established in 1893.

In fact, no independent Afghan government ever accepted it. But we're demanding that Pakistan block any Pashtun effort to overturn what they have never accepted and Afghanistan has never accepted. We're driving Pakistanis to a very dangerous position. One of the interesting WikiLeaks exposures was from Anne W. Patterson, the American ambassador in Pakistan, who supports U.S. policy in Pakistan but pointed out that it carries with it the danger of "destabilizing the Pakistani state," maybe even leading to a coup, which could bring about the leaking of radioactive materials into the jihadi networks.[36] The jihadists are not the dominant force in Pakistan, but they're present, and have been since the radical Islamization during the Reagan years. Reagan and Saudi Arabia were supporting the worst dictator in Pakistan's history, Muhammad Zia-ul-Haq. One of his primary goals was to bring about a radical Islamization of the country, establishing madrassas all over the place. That's where the Taliban come from.

So yes, there is a radical Islamic element in Pakistan, and it's almost certainly engaged in some fashion in the vast nuclear industry. It's conceivable that under pressure you might find leakage of nuclear materials to jihadi hands, which could lead to a dirty bomb in London or New York. It's likely.

6

MENTAL SLAVERY

CAMBRIDGE, MASSACHUSETTS (JANUARY 20, 2012)

Bob Marley, the famous reggae singer from Jamaica, sang a popular lyric: "Emancipate yourself from mental slavery."[1] *That's a theme that you've returned to quite a bit in your work.*

I should know that song. Yes, it's true. When people wanted enough freedom that they couldn't be enslaved or killed or repressed, new modes of control naturally developed to try to impose forms of mental slavery so they would accept a framework of indoctrination and wouldn't raise any questions. If you can trap people into not noticing, let alone questioning, crucial doctrines, they're enslaved. They'll essentially follow orders as if there was a gun pointed at them.

In some of your talks, when people ask you what to do in response to the problems you discuss, I have heard you tell people they could start by turning off their television set.

Television drums certain fixed boundaries of thought into your head, which certainly dulls the mind. The doctrines are not formally stated. It's not the Catholic Church: "You have to believe this. You have to read this every day, say this every day." It's just presupposed. You presuppose a framework, and then people just come to accept it.

A decent propaganda system does not announce its principles or intentions. This is one of the reasons the old Soviet system was relatively ineffective, as far as we know. If you tell people, "This is what you have to think," then they understand: this is what power wants us to think. And then they may find a way out of it. It's harder to extricate yourself from a system of unstated presuppositions than it is from explicitly stated doctrine. That's the way a good propaganda system will operate.

Our propaganda system is highly sophisticated. The actors substantially understand what they're doing, it seems. Take the 2008 presidential election, which, like all elections, was a public relations extravaganza. The advertising industry was very conscious of its role. In fact, shortly after the election, *Advertising Age* gave the annual prize for best marketing campaign of the year to the Obama campaign, which the PR industry organized.[2]

There was actually discussion in the business press afterward over this achievement.[3] There was euphoria in the business community. This will change the style in corporate boardrooms. We know how to delude people better than we did before. No one had any illusions, apparently, about the candidate winning on the basis of his policies or his intentions. It was just a good marketing campaign, better than John McCain's.

In an image-dominated culture, I wonder about the future of books. And I'm asking this of someone who reads voraciously. Your reading habits are legendary. We're sitting in your office, surrounded by piles and piles of books. How do you get through all this stuff?

Unfortunately, I don't. This is the urgent pile. There are many more stacks elsewhere. But one of the painful experiences which I try to avoid as much as possible is to calculate how much time it would take, if I read constantly, to go through them. And reading a book doesn't just mean turning the pages. It means thinking about it, identifying parts that you want to go back to, asking how to place it in a broader context, pursuing the ideas. There's no point in reading a book if you let it pass before your eyes and then forget about it ten minutes later. Reading a book is an intellectual exercise, which stimulates thought, questions, imagination.

I suspect that will disappear. You see various signs of

it. There has been a shift in my own classes over the past ten or twenty years. Whereas I once could make casual literary references and people more or less knew what I was talking about, this is less and less true. I can see from correspondence that people are constantly asking questions about something they saw on YouTube but not about an article or a book. They very often rightly ask, "You said so-and-so. What's the evidence for it?" In fact, in an article I wrote the same week as that talk, there might have been footnotes and discussion, but it doesn't occur to them to look for that.

What does that mean for an intellectual culture, then?

It's going to degrade the intellectual culture. It can't help but do so. It's a mixed story. Take, say, electronic books. They have advantages. You have half a dozen books you can read on an airplane trip. On the other hand, when I read a book I care about, I want to make comments in the margins, I want to underline things. I want to make notes on the flyleaf. Otherwise I don't even know what to go back to. You can't do that the same way with an electronic book. Words just pass into your eyes. Maybe they don't even stay in your brain.

The same is true of the Internet. Access to the Internet is a great thing. A huge amount of material is available. On the other hand, it's evanescent. Unless you know what you're looking for, and you store it properly and put it into context, it's as if you never saw it. There's no point in having a lot of data available unless you can make some sense out of it. And that takes thought, reflection, inquiry. I think these capacities are being degraded to an extent. You can't measure it, but I have a sense that's true.

What do you think about Twitter? You have one hundred and forty characters to express something.

Yes. Bev Stohl, who works with me at MIT, told me about it. I get a ton of e-mail. Increasingly, the messages I receive have been one-sentence queries or comments, sometimes so brief that they're in the subject line of the e-mail. Bev pointed out to me that those are the length of Twitter messages. If you look at them, they have a fairly consistent character. They give the impression of being something that someone just thought of. You're walking down the street, a thought comes to you, you tweet it. If you thought for two minutes, or if you had made the slight effort involved in looking up the topic, you wouldn't have sent it. In fact, it has reached the point that sometimes I just send a form letter saying I can't respond to a one-line question.

Getting back to books, your lectures are replete with references to information that you learned in books, for example, something about Martin Luther King that Taylor Branch wrote or something about the U.S. labor movement by David Montgomery.

You're able to bring this reading knowledge into the intellectual formulations you then present.

Anybody can do it. It's not a special talent. But you have to be willing to think about what you're reading. You can be led down a false track. You can be deluded. The same is true in the sciences. You can be pursuing some idea that you really think is exciting, work hard on it, get what looks like an explanation, and then you find out you were going in the wrong direction and you have to backtrack. You can learn from that, too. But if you don't stop to think, reflect, and find a context, it's wasted effort. You might as well not be reading.

I was struck in a talk in New York that you mentioned E. L. Doctorow's Ragtime.[4] *Was that the last novel you read?*

I think the last novel I read was by the Icelandic Nobel Prize laureate Halldór Laxness. I was in Iceland. Somebody lent me one of his novels, and I read it on the plane back. It's great. When I was in England about a year ago, a friend gave me *A Case of Exploding Mangoes*, a Pakistani novel by Mohammed Hanif.[5] It was very good. I can't read as much fiction as I'd like.

States around the world, from China to Syria to the United States, are becoming increasingly nervous about the Internet and social media. Calls for control and censorship of the Internet are increasing.

Right now there is a big battle going on among the titans of industry over new proposed legislation called the Stop Online Piracy Act. The movie industry, the record industry, and other big operators want to restrain what they call piracy, people taking their products without payment or agreement. But there are other big corporations that are pushing back. Wikipedia shut down for a day in protest.[6] Google, one of the biggest corporations in the world, also protested.[7]

Every rich and developed country carried out piracy. During its period of rapid growth, the United States stole more efficient advanced British technology. Britain did the same with countries that it was crushing: Ireland, the Low Countries, Belgium, India. It's what we accuse China of doing today, following in our footsteps.

The trade agreements imposed by the rich and powerful level very heavy penalties against piracy. So-called intellectual property rights are built into the World Trade Organization rules and other trade agreements, with very stringent requirements. One of the most important examples is the protection given to the pharmaceutical industry. So, for example, there are guidelines to prevent countries with pharmaceutical industries, like India, from producing cheap drugs that will be available to the general population, undercutting the profits of major international corporations.

The pharmaceutical companies argue they need the profits for new research and development. Otherwise there won't be new drugs. The movie and record industries say they need massive profits to support creative artists. These arguments have a superficial plausibility until you look into the issue closely. The economist Dean Baker has shown pretty conclusively that these are not persuasive arguments. So, say, with pharmaceutical research and development, according to his calculations, which look pretty reasonable to me, if the pharmaceutical companies were forced on the market and the entire research and development cost were picked up by the public, there would be a huge saving to the public, because most of the work is done under public auspices anyway, at universities, the National Institutes of Health.[8] The pharmaceutical companies pick it up at the sort of tail end and do the testing, the marketing, the packaging. So yes, they make a contribution, but a lot of their effort is put into making copycat drugs. You just shift a molecule so you can sell something.

With regard to creative artists, Baker also has some suggestions that seem sensible to me. Namely, they should be publicly funded.[9] That's basically what happens with, say, classical music or opera. If you could extend that, you wouldn't need intellectual property rights and the piracy issue would disappear.

How does the United States square its trumpeting of the free flow of information and democratic rights of expression with its response to WikiLeaks?

The profession of dedication to rights is always tinged with a fundamental hypocrisy: rights if we want them, not if we don't. The clearest example of this is support for democracy. It's pretty well established over many decades that the United States supports democracy only if it accords with strategic and economic objectives. Otherwise it opposes it. The United States is by no means alone on that, of course. The same is true of terror, aggression, torture, human rights, freedom of speech, whatever it might be.

So the line that the enormous trove of information that was disseminated through WikiLeaks was somehow compromising U.S. security doesn't wash.

It compromised the security that governments are usually concerned about: their security from inspection by their own populations. I haven't read everything on WikiLeaks, but I'm sure there are people who are searching very hard to find some case where they can claim there has been a harm to genuine security interests. I couldn't find any myself.

One respect in which the United States is unusually open is in declassifying government documents. By comparative standards we have better access to internal government decisions than any country that I know of. The system isn't perfect, but there's a regular declassification procedure—the Freedom of Information Act functions to an extent—and there is a fair amount of access. I've spent a lot of time working through declassified documents, and most of them are just totally boring. You can read through volume after volume of the *Foreign Relations of the United States* and maybe you'll find three sentences that are worth paying attention to. Many of the classified documents have little to do with genuine security but a lot to do with preventing the population from knowing what the government is up to. I think that's been true of what I've seen of WikiLeaks, too.

Take the one example I mentioned, Ambassador Patterson's comments about Pakistan and the danger of the Bush-Obama policy destabilizing a country with one of the biggest nuclear weapons programs in the world, in fact, one that's growing fast and interlaced with jihadi elements. That's something the population ought to know about, but it has to be kept from them. You have to describe our policies in terms of defending ourselves from attack when you're in fact increasing the threat of attack. That's true over and over again.

There are other interesting WikiLeaks exposures. At the time of the military coup in Honduras in 2009, the embassy in Honduras carried out an extensive investigation to determine whether the coup was legal or illegal, and they concluded, "The Embassy perspective is that there is no doubt that the military, Supreme Court and National Congress conspired on June 28 in what constituted an illegal and unconstitutional coup against the Executive Branch."[10] That assessment was sent back to Washington, which means the Obama administration knew about it, but they discarded the finding and, after various steps, ended up supporting the military coup, as they still do.[11] For people who want to understand Obama's thinking about freedom and democracy, that's important information. But it's not something the government wants you to know.

Actually, one of the most interesting aspects of the WikiLeaks exposures is how they were treated. Some of the exposures were heralded as a wonderful event. For example, there were exposures related to diplomatic cables. These are diplomatic cables, so you don't know how accurate they are. Diplomats tend to report what they know the center wants to hear, so filtering is already taking place. But there were cables from the Middle East embassies saying that the Arab dictators support U.S. policy on Iran. The king of Saudi Arabia was quoted as saying that we have "cut off the head of the snake."[12] That was all over the headlines. There were articles by leading commentators, such as Jacob Heilbrunn, saying that this is fantastic.[13] WikiLeaks should be congratulated for showing us how marvelous we are that the Arab dictators support us. It's as if the CIA is running WikiLeaks.

At the same time this discussion was happening about how the Arab world supports U.S. goals in Iran, a U.S.-run poll showing that the Arab populations are strongly against U.S. policy in Iran was released. So strongly against it that in, say, Egypt about 80 percent of the population thought the region would be more secure if Iran had nuclear weapons.[14] They're concerned with the real threats, the United States and Israel.[15] But that material was barely published. So here we have applause for the fact that the dictators support us, silence about the fact that the populations strongly oppose us. That tells you something about our commitment to democracy.

There are indications that the cables WikiLeaks exposed about the dictatorship of Zine El-Abidine Ben Ali in Tunisia had a big influence on the revolt there.

It's questionable. The leaks showed that the U.S. government understood very well that Ben Ali was a harsh, corrupt dictator, that the population was very upset and strongly opposed to him.[16] But that had no effect on support for his regime.

Do you mean support from Washington?

U.S. support. French support, primarily. France was just outlandish. After the uprising had already started, one French cabinet minister, Michèle Alliot-Marie, actually went to Tunisia for a vacation.[17] This is a country that's been under the thumb of France for a long time and is surely penetrated by French intelligence. But how much these leaks influenced the protests is an open question. I doubt that Tunisians cared very much about French and U.S. hypocrisy, which is all that WikiLeaks revealed—nothing that they didn't know themselves.

Talk about the connection between Daniel Ellsberg and Bradley Manning.

Dan is an old friend. I was involved with him in helping release the Pentagon Papers, which I thought was a quite proper thing to do. I testified at his trial. In the case of Bradley Manning, he's charged with having released material to Julian Assange, who distributed it on WikiLeaks.[18] He's been in prison now since May 2010, a large part of that in solitary confinement—which is torture. He's been treated in rotten ways and been bitterly attacked.

Here's someone who is charged with doing something which, in my opinion, is not a crime but a service to the country. But whatever you think about that, he's charged, not brought to trial. In fact, at the moment there's not even any court trial contemplated. They're treating it as a court-martial inside the military system.[19]

I think Manning should be applauded and the government should be harshly condemned for throwing out the basic principles of law and human rights.

Didn't Obama, a constitutional law professor, make a prejudicial comment about Bradley Manning?

Yes, he immediately said he's guilty.[20] That's unconscionable. Even if Obama wasn't a constitutional lawyer, he's the president. He should know that the president shouldn't say that about a person who is facing criminal charges.

There are worse things—say, assassinating Osama bin Laden. He wasn't tried in a court. He's innocent until proven guilty. But you assassinate him if you don't like him.

As they also did to Anwar al-Awlaki in Yemen, a U.S. citizen.[21]

That case got a little attention because al-Awlaki is a U.S. citizen. Maybe he's guilty of something, maybe not. But if, say, Iranian terrorists killed somebody tomorrow—say, Leon Panetta, the defense secretary—because he's involved in planning attacks against Iran, which he is, would we think that's okay?

It seems that many liberals who criticized war crimes during the presidency of George W. Bush have been relatively muted during the Obama period.

They have. Some people are talking, but not many. Obama has also made it clear that nobody is going to be punished for war crimes in the Bush period, which is quite understandable.[22] If people were punished for that, then he could be punished for similar crimes today.

That brings me to a comment that you made years back that every president since 1945 could be tried for war crimes.[23] Do you still hold to that?

I think I was pretty careful. I said that would be a fair statement by the principles of Nuremberg. Not by the practice of Nuremberg, which departed sharply from the principles.

The principle being that "planning, preparation, initiation or waging of a war of aggression or a war in violation of international treaties, agreements or assurances" is an international war crime.[24]

That was the primary charge, but there were many others. So, for example, one of the main charges against Joachim von Ribbentrop, the German foreign minister, who was hanged after the war, was that he either permitted or was complicit in a preemptive strike against Norway. Norway really posed a threat to Germany. The British were there and were planning an attack on Germany. Compare this to what happened to Colin Powell when he was complicit in a preemptive strike against Iraq. Powell was not tried for having gone to the United Nations and producing fabricated stories calling for an attack on Iraq, where there was no threat at all, in fact not even a remote threat.

So there are the principles of Nuremberg. But, of course, the practical outcome is quite different. The Nuremberg tribunal was the most authentic and significant of any of the international war crimes tribunals that have taken place, but it had fundamental flaws. And they were known to the prosecutors. For example, Telford Taylor commented on them right away. Effectively, he said, the tribunal defined war crimes as something you did and we didn't do.[25] That was the criterion. So, for example, the bombing of civilian concentrations, urban bombing, was not considered a war crime because the Allies did much more of it than the Axis. In fact, German admiral Karl Dönitz was able to reject the charges against him because he got testimony from the British admiralty and from the American navy saying we did the same thing, so it definitely isn't a war crime.[26]

One of the things that you say about yourself, which often stuns people, is that you're an old-fashioned conservative. What do you mean by that?

For example, I think Magna Carta and the whole legal tradition that grew out of it made some sense. I think the expansion of the moral horizon over the centuries, particularly since the Enlightenment, is important.

I think there's nothing wrong with those ideals. A conservative, at least as it used to be understood, is somebody who cares about traditional values. Today those values are regularly being thrown out the window. We should condemn that.

Then why are you seen as a wild-eyed radical?

Because holding on to traditional values is a very radical position. It threatens and undermines power.

A perennial question that you get at your talks is, "Well, there's an election coming up, Professor Chomsky. What should I do? Should I vote? Should I stay at home?"

The first point is, I think you should spend about five minutes on the question. There are much more important questions, such as, "What should I be doing to try to change the country?" But the question about elections doesn't take much thought, in my opinion. When we get to the presidential election—let's put aside the primaries—you're going to have a small range of choices. There will be two candidates, neither of whom you like. One will be plausibly much more dangerous than the other. If you're in a so-called safe state, where you know how the vote is going to come out, you have choices. You can say, "Okay, I won't vote—or I'll vote for some party that's trying to become an independent alternative, say, the Greens." If you're in a swing state, you have to ask yourself, "Do I want to help the worse candidate be elected or do I want to prevent that?" It doesn't mean you like the other candidate. But, in fact, that is the choice. So you have to ask, "Is it better to help the worse candidate to be elected?" You can make a case for doing that. In fact, there was an old Communist Party principle back in the early 1930s: "the worse, the better." If you get the worse candidate in, it's going to be better, because then there will be more support for a revolution. That was the choice some people made in Germany, and you know where that led. So it's a question you have to think about, but I don't think you have to think about it very hard.

Do you think the Occupy movements should be involved in electoral politics or work from below without engaging in the system?

As they stand, they're not an electoral force. First of all, I don't think they can take a unified position. They have no mechanism for making a unified decision, and I think that's a good thing. It's better to have a variety of opinions and attitudes, as well as interchange and interaction about what to do, and to accept and tolerate opposing opinions within a general framework. I think that's much more important than having a general assembly vote saying we support X or Y or Z.

What are some practical steps that you think the movements can undertake?

They've already undertaken practical steps. So, for example, they have substantially changed the general discourse in the country. There is now overt concern and engagement in questions of inequality, the extraordinary power of financial institutions, government subordination to financial institutions, the role of finance and of money in general in the buying and shaping of elections. And they can go further—and already have, to some extent. So, for example, they can ask: Why should it be up to executives and managers to decide to settle investment decisions about where things will be produced and what will be produced, how profits will be distributed? Why should that be the domain of the directorate of a corporation? Usually a bank is a small sector of rich people. Do they have some natural right to make those decisions? Not by any economic principle. In fact, there's every reason to advocate that those decisions be made by what are called stakeholders—communities, the workforce, others affected by what's decided.

But going forward, how are they going to be able to sustain themselves in the face of this propaganda system and an increasingly repressive police force? One of the things that a number of people have commented on has been the degree of militarization of local police departments.[27] *They're looking more and more like special operations forces.*

Power doesn't commit suicide. So yes, there will be attempts to carry out repression. But the repression that exists now is not remotely like it has been in the past. There's nothing like Wilson's Red Scare or COINTELPRO. As far as we are aware, there are no assassinations of movement leaders. But, yes, there is repression. And some of the tactics Occupy has used, which are good tactics, lend themselves easily to police repression. So occupying a space is a very good tactic. I think it's good that Occupy has done it. But you have to recognize that it provides an opening to police attacks, which probably would be publicly supported to a large extent. So you need to find other tactics.

The way to deal with the repression and denigration that will take place is to build more popular support. What the Occupy movements have to do if they're going to sustain themselves is recognize that tactics are not strategies. A tactic may be a very good one, but tactics tend to have diminishing returns after a while. People get tired of them and they lose their efficacy, so you have to move on. I think it's generally accepted in the movements that they have to reach out and engage other sectors of the society. There have been moves in that direction, like joining with anti-foreclosure movements. But again, active labor participation will be essential.

Let's talk more about the environment. You say that "risk in the financial system can be remedied by the taxpayer, but no one will come to the rescue if the environment is destroyed. That it must be destroyed is close to an institutional imperative."[28] *Explain.*

It is an institutional imperative. By imperative, I don't mean it's a law of nature. You can change it. But given the way institutions now function, their core goal is to maximize short-term profit and power. That is a critical element for the core of decision makers in the economy and the society—and, therefore, in the political system. And that leads almost directly to destruction of the environment. In fact, we can see it right in front of us. The threat is quite serious. The major agencies that monitor global emissions have released very ominous predictions. The International Energy Association (IEA) released data which their own chief economists concluded give us maybe five more years before we reach a turning point that will be irreversible.[29]

Fatih Birol, chief IEA economist, has said, "The door is closing. . . . I am very worried—if we don't change direction now on how we use energy, we will end up beyond what scientists tell us is the minimum [for safety]. The door will be closed forever."[30]

The IEA is a pretty conservative agency. This is not a bunch of radicals. In fact, it was formed through the initiative of

Henry Kissinger. I haven't seen much reporting about it, but one of the few news articles quoted John Reilly, the codirector of MIT's Joint Program on the Science and Policy of Global Change, who also said that the IPCC estimates were too low.[31] "The more we talk about the need to control emissions, the more they are growing," he warned, and if we don't do something very quickly about fossil fuels we're going to be over the edge. "Increasing reliance on coal is imperiling the world," he added. Again, this is not coming from far-out radicals but from major institutions, leading scientists.

It's interesting to watch the way climate change is discussed in the media. It's usually presented as a he-says-she-says issue. On the one hand, you have the IPCC. On the other hand, you have a handful of scientists and a couple of senators who say, "We don't believe any of it." That's the choice. Actually, there is a third set of scientists, who almost never make it into print, and it's much larger than the fringe of denialists: people who say that the consensus is much too conservative, that the risks are much higher. People like the ones who run the MIT program I mentioned or the chief economist of the International Energy Association. But they are ignored, and we almost never hear their views. And the public is left with a choice between two positions, which they're in no position to make a judgment about.

On top of that, you have a huge propaganda offensive

from the business sector, saying, "Don't believe any of it. None of it is real." A little to my surprise, this has even affected the more serious and responsible parts of the business press, like the *Financial Times,* maybe the best newspaper in the world. Just at the time that these emissions reports were coming out, the *Financial Times* euphorically suggested that the United States was entering a new age of plenty and might have a century of energy independence, even global hegemony, ahead of it thanks to the new techniques of extracting fossil fuels from shale rock and tar sands.[32] Leaving aside the debates about whether these predictions are right or wrong, celebrating this prospect is like saying, "Fine, let's commit suicide." I'm sure the people who write such articles have read the same climate change reports I have and take them seriously. But their institutional role makes such positions a social or cultural necessity. They could make different decisions, but that would require real rethinking of the nature of our institutions.

The propaganda barrage has been effective. As Naomi Klein writes in the Nation, *"A 2007 Harris poll found that 71 percent of Americans believed that the continued burning of fossil fuels would cause the climate to change. By 2009 the figure had dropped to 51 percent. In June 2011 the number of Americans who agreed was down to 44 percent—well under half the population. According to Scott Keeter, director of survey research at the Pew Research Center for People and the Press, this is 'among the largest shifts over a short period of time seen in recent public opinion history.'* "[33]

A significant majority of Americans still think climate change is a serious problem, but it's true that it has declined. The Pew polls are quite interesting in that they're international polls, and they show that internationally there's very strong concern. The United States is not totally off the spectrum, but it's close to the edge. Concern in the United States is notably less than in comparable countries. And the drop that Klein is describing is exactly what they report. It's very hard to doubt that that's connected with the propaganda campaign that has been quite openly conducted.

In fact, a couple of years ago, right after the insurance company victories on the health reform bill, so-called Obamacare, there was a report in the *New York Times* about leaders of the American Petroleum Institute and other business groups looking to the victory in the health care campaign as a model to undermine concern about global warming.[34] In the Republican presidential debates, for example, even to mention global warming would be to commit political suicide.

Some of the candidates have remarkable positions on climate change. Take Ron Paul. He appeals to a lot of progressives. He said on Fox, "The greatest hoax I think that has been around for many, many years if not hundreds of years has been this hoax on the environment and global warming."[35] He doesn't provide any argument or evidence as to why he disregards the scientific consensus—just, I say so, period. With that attitude, you really are approaching the edge.

And, in fact, actions are being taken to implement those views. A sign of the shift in the nature of elite discourse in recent years is that the Republicans in Congress are now trying to dismantle the few environmental regulations and controls that do exist, which were instituted under Nixon. Nixon would look like a radical today, Dwight Eisenhower like a super radical.

7

LEARNING HOW TO DISCOVER

CAMBRIDGE, MASSACHUSETTS (MAY 15, 2012)

It's been more than five decades since you first wrote about universal grammar, the idea of an inborn capacity in every human brain that allows a child to learn language. What are some of the more recent developments in the field?

Well, that gets technical, but there's very exciting work going on refining the proposed principles of universal grammar. The concept is widely misunderstood in the media and in public discussions. Universal grammar is something different: it is not a set of universal observations about language. In fact, there are interesting generalizations about language that are worth studying, but universal grammar is the study of the genetic basis for language, the genetic basis of the language faculty. There can't be any serious doubt that something like that exists. Otherwise an infant couldn't reflexively acquire language from whatever complex data is around. So that's not controversial. The only question is what the genetic basis of the language faculty is.

Here there are some things that we can be pretty confident about. For one thing, it doesn't appear that there's any detectable variation among humans. They all seem to have the same capacity. There are individual differences, as there are with everything, but no real group differences—except maybe way at the margins. So that means, for example, if an infant from a Papua New Guinea tribe that hasn't had contact with other humans for thirty thousand years comes to Boulder, Colorado, it will speak like any kid in Colorado, because all children have the same language capacity. And the converse is true. This is distinctly human. There is nothing remotely like it among other organisms. What explains this?

Well, if you go back fifty years, the proposals that were made when this topic came on the agenda were quite complex. In order just to account for the descriptive facts that you saw in many different languages, it seemed necessary to assume that universal grammar permitted highly intricate mechanisms, varying a lot from language to language, because languages looked very different from one another.

Over the past fifty to sixty years, one of the most significant developments, I think, is a steady move, continuing today, toward trying to reduce and refine the assumptions so that they maintain or even expand their explanatory power for particular languages but become more feasible with regard to other conditions that the answer must meet.

Whatever it is in our brain that generates language developed quite recently in evolutionary time, presumably within the last one hundred thousand years. Something very significant happened, which is presumably the source of human creative endeavor in a wide range of fields: creative arts, tool making, complex social structures. Paleoanthropologists sometimes call it "the great leap forward." It's generally assumed, plausibly, that this change had to do with the emergence of language, for which there's no real evidence before in human history or in any other species. Whatever happened had to be pretty simple, because that's a very short time span for evolutionary changes to take place.

The goal of the study of universal grammar is to try to show that there is indeed something quite simple that can meet these various conditions. A plausible theory has to account for the variety of languages and the detail that you see in the surface study of languages—and, at the same time, be simple enough to explain how language could have emerged very quickly, through some small mutation of the brain, or something like that. There has been a lot of progress toward that goal and, in a parallel effort, to try to account for the apparent variability of languages by showing that, in fact, the perceived differences are superficial.

The seeming variability has to do with minor changes in a few of the structural principles that are fixed.

Discoveries in biology have encouraged this line of thinking. If you go back to the late 1970s, François Jacob argued that it could well turn out—and probably is true—that the differences between species, let's say an elephant and a fly, could be traceable to minor changes in the regulatory circuits of the genetic system, the genes that determine what other genes do in particular places. He shared the Nobel Prize for early work on this topic.

It looks like something similar may be true of language. There's now work on an extraordinarily broad range of typologically different languages—and, more and more, it looks like that. There's plenty of work to do, but a lot of this research falls into place in ways that were unimaginable thirty or forty years ago.

In biology it was plausible quite recently to claim that organisms can vary virtually without limit and that each one has to be studied on its own. Nowadays that has changed so radically that serious biologists propose that there's basically one multicellular animal—the "universal genome"—and that the genomes of all the multicellular animals that have developed since the Cambrian explosion half a billion years ago are just modifications of a single pattern. This thesis hasn't been proven, but it is taken seriously.

Something similar is going on, I think, in the study of language. Actually, I should make it clear that this is a minority view, if you count noses. Most of the work on language doesn't even comprehend these developments or take them seriously.

Is the acquisition of language biological?

I don't see how anyone could doubt that. Just consider a newborn infant. The newborn is barraged by all kinds of stimuli, what William James famously called "one great blooming, buzzing confusion."[1] If you put, say, a chimpanzee or a kitten or a songbird in that environment, it can only pick out what's related to its own genetic capacities. A songbird will pick out a melody of its species or something from all this mass because it's designed to do that, but it can't pick out anything that's relevant to human language. On the other hand, an infant does. The infant instantly picks language-related data out of this mass of confusion. In fact, we now know that this goes on even in the uterus. Newborn infants can detect properties of their mother's language as distinct from certain—not all, but certain—other languages.

And then comes a very steady progression of acquisition of complex knowledge, most of it completely reflexive. Teaching doesn't make any difference. An infant is just picking it out of the environment. And it happens very fast, in a very regular fashion. A lot is known about this process. By about six months, the infant has already analyzed what's called the prosodic structure of the language, stress, pitch—languages differ that way—and has sort of picked out the language of its mother or whatever it hears, its mother and its peers. By about nine months, roughly, the child has picked out the relevant sound structure of the language. So when we listen to Japanese speakers speaking English, we notice that, from our point of view, they confuse "r" and "l," meaning they don't know the distinction. That's already fixed in an infant's mind by less than a year old.

Words are learned very early, and, if you look at the meaning of a word with any care, it's extremely intricate. But children pick up words often after only one exposure, which means the structure has got to be in the mind already. Something is being tagged with a particular sound. By, say, two years, there's pretty good evidence that the children have mastered the rudiments of the language. They may just produce one-word or two-word sentences, but there's now experimental and other evidence that a lot more is in there. By three or four, a normal child will have extensive language capacity.

Either this is a miracle or it's biologically driven. There are just no other choices. There are attempts to claim that language acquisition is a matter of pattern recognition or memorization, but even a superficial look at those proposals shows that they collapse very quickly. It doesn't mean that they're not being pursued. In fact, those lines of inquiry are very popular. In my view, though, they're just an utter waste of time.

There are some very strange ideas out there. For instance, a lot of quite fashionable work claims that children acquire language because humans have the capacity to understand the perspective of another person, according to what's called theory of mind. The capacity to tell that another person is intending to do something develops in normal children at roughly age three or four. But, in fact, if you look at the autism spectrum, one of the classic syndromes is failure to develop theory of mind. That's why autistic kids, or adults for that matter, don't seem to understand what other people's intentions are. Nevertheless, their language can be absolutely perfect. Furthermore, this capacity to understand the intention of others develops long after the child has mastered almost all the basic character of the language, maybe all of it. So that can't be the explanation.

There are other proposals which also just can't be true, but are still pursued very actively. You read about them in the press, just as you read things about other organisms having language capacity. There's a lot of mythology about language, which is very popular. I really don't want to sound too dismissive, but I feel dismissive. I think these ideas can't be considered seriously.

Whatever our language faculty is, humans develop it very quickly, on very little data. In some domains, like the meaning of expressions, there's virtually no data. Nevertheless it's picked up very quickly and very precisely, in complex ways. Even with sound structure, where there's a lot of data—there are sounds around, you hear them—it's still a regular process and it's distinctively human. Which is striking, because it's now known that the auditory systems of higher apes, say chimpanzees, appear to be very similar to the human auditory system, even picking out the kinds of sounds that play a distinctive role in human language. Nevertheless, it's just noise for the ape—they can't do anything with it. They don't have the analytical capacities, whatever they are.

What's the biological basis for these human capacities? That's a very difficult problem. We know a lot, for example, about the human visual system, partly through experimentation. At the neural level, we know about it primarily from invasive experiments with other species. If you conduct invasive experiments on other mammals, cats or monkeys, you can find the actual neurons in the visual system that are responding to a light moving in a certain direction. But you can't do that with language. There is no comparative evidence, because other species don't have the capacity and you can't do invasive experiments with humans. Therefore, you have to find much more complex and sophisticated ways to try to tease out some evidence about how the brain is handling all this. There's been some progress in this extremely difficult problem, but it's very far from yielding the kind of information you could get from experimentation.

If you could experiment with humans, say, isolating a child and controlling carefully the data presented to it, you could learn quite a lot about language. But obviously you can't do that. The closest we've come is looking at children with sensory deprivation, blind children, for example. What you find is pretty amazing. For example, a very careful study of the language of the blind found that the blind understand the visual words *look, see, glare, gaze,* and so on quite precisely, even though they have zero visual experience. That's astonishing. The most extreme case is actually material that my wife, Carol, worked on, adults who were both deaf and blind. There are techniques for teaching language to the deaf-blind. Actually, Helen Keller, who is the most famous case, invented them for herself. It involves putting your hand on somebody's face, with your fingers on the cheeks and thumb on the vocal cords. You get some data from that, which is extremely limited. But that's the data available to the deaf-blind, and they have pretty remarkable language capacity. Helen Keller was incredible, a great writer, very lucid. She's an extreme case.

Carol did a study here at MIT. She found in working with people with sensory deprivation that they achieved pretty remarkable language capacity. You have to do quite subtle experiments to find things they don't know. In fact, they managed to get along by themselves. The primary subject, the one most advanced, was a man who was a tool and die maker, I think. He worked in a factory somewhere in the Midwest. He lived with his wife, who was also deaf-blind, but they found ways to communicate with buzzers in the house and things that you could touch that vibrated. He was able to get from his house to Boston for the experiments by himself. He carried a little card which said on it, "I am deaf-blind. May I put my hand on your face?" so, if he got lost, if somebody would let him do that, he could communicate with them. And he lived a pretty normal life.

One very striking fact was that all of the cases that succeeded were people who had lost their sight and hearing at about eighteen months old or older—it was primarily through spinal meningitis in those days. People who were younger than that when they became deaf-blind never learned language. There weren't enough cases to actually prove anything, so the results of the study were never published, but this was a pretty general result. Helen Keller fits. She was twenty months old when she lost her sight and hearing. It suggests, at least, that by eighteen or twenty months, a tremendous amount of language is already known. It can't be exhibited but it's in there somewhere, and can possibly be teased out later.

It's known that the ability to acquire language starts decreasing rather sharply by about the mid-teens.

That's descriptively correct, although, again, it's not 100 percent correct. There is individual variation. There are individuals who can pick up a language virtually natively at a much later age. Actually, one of them was in our department. Kenneth Hale, one of the great modern linguists, could learn a language like a baby. We used to tease him that he just never matured.

That's an exception?

Yes. By and large, what you said is true. The basis is not really known, but there are some thoughts about it. One thing we know is that, from the very beginning, brain development entails losing capacities. Your brain is originally set up so that it can acquire anything that a human can acquire. In the case of language, say, it's set up so that you can acquire Japanese, Bantu, Mohawk, English, whatever. Over time that declines. In some cases, it declines even after a few months of age. What's happening across all cognitive capacities, not only in the case of language, is that synaptic connections, connections inside the brain, are being lost. The brain is being simplified, it's being refined. Certain things are becoming more effective, other things are just gone. There's apparently a lot of synaptic loss around the period of puberty or shortly beforehand, and that could be relevant.

I attended one of your seminars in linguistics here at MIT a few years ago, and I was struck by a couple of things. First of all, I was one of the few non-Asians in your class. It was mostly

South Asians and East Asians. But the other thing was the extent to which math was involved. You were constantly writing formulas on the blackboard.

We should be clear about that. It's not deep mathematics. It's not like proving hard theorems in algebraic topology or something. But there's good reason why some sophistication in mathematics is at least advantageous, maybe necessary, for advanced work. The basic reason is that language is a computational system. So whatever else it is, the capacity we're both using and sharing is based on a computational procedure that forms an infinite array of hierarchically structured expressions.

A lot of people conflate linguistics with the ability to speak many languages. So in your case, people think, Oh, Chomsky, he must know ten or twelve languages. But in fact linguistics is another universe. Explain why the study of language is important. Clearly, you're animated by it. You've devoted most of your life to it.

I should say, sometimes there's a distinction made between languist and linguist. A languist is somebody who can speak a lot of languages. A linguist is somebody who is interested in the nature of language.

Why is it interesting? Think about the picture that I presented before, which I think is fairly uncontroversial. At some time in the very recent past, from an evolutionary point of view, something quite dramatic happened in the human lineage. Humans developed what we now have: a very wide range of creative capacities that are unknown in the previous record or among other animals. There is no analogue to them. That's the core of human cognitive, moral, aesthetic nature—and right at the heart of it was the emergence of language.

In fact, it's very likely that language was the lever by which the other capacities develop. In fact, other capacities may just be piggybacking off language. It's possible that our arithmetical capacities and—quite likely—our moral capacities developed in a comparable way, maybe drawing from the analytical, computational mechanisms that yield language in all of its rich complexity. To the extent that we understand these other things, which is not very much, it seems that they're using the same or similar computational mechanisms.

Clearly, culture influences and shapes language, even if it doesn't determine it.

That's a common comment, but it's almost meaningless. What's culture? Culture is just a general term for everything that goes on. Yes, sure, everything that goes on influences language.

If we're, let's say, in a violent environment, doesn't that shape the vocabulary? Wouldn't that lead us to talk about "epicen-

ter" and "Ground Zero" and "terrorism" and other terms in the lexicon of violence?

Sure, there's an effect on lexical choices. But that's peripheral to language. You could take any language that exists and add those concepts to it—a fairly trivial matter. But we don't know anything really about the effects of culture on lexical choices. In my view, it's unlikely cultural environments meaningfully affect the nature of language. Take, say, English, and trace it back to earlier periods. English was different in Chaucer's time or King Arthur's time, but the language hasn't fundamentally changed, the vocabulary has. Not long ago Japan was a feudal society, and now it's a modern technological society. The Japanese language has changed, of course, but not in ways that reflect those changes. And if Japan went back to being a feudal society, the language wouldn't change much either.

Vocabulary does, of course. You talk about different things. For example, the tribe in Papua New Guinea that I mentioned before wouldn't have a word for computer. But again that's fairly trivial. You could add the word for computer. Ken Hale's work from the 1970s on this question is quite instructive. He was a specialist on Australian aboriginal languages, and he showed that many of these languages appear to lack elements that are common in the modern Indo-European languages. For example, they don't have words for numbers or colors and they don't have embedded relative clauses. He studied this topic in depth and showed that these gaps were quite superficial. So, for example, the tribes he was working with didn't have numbers, but they had absolutely no problem counting. As soon as they moved into a market society and had to deal with counting, they just used other mechanisms. Instead of number words, they would use their hand for *five*, two hands for *ten*. They didn't have color words. Maybe they just had *black* and *white*, which apparently every language has. But they used expressions such as *like blood* for what we would call *red*.

Hale's conclusion was that languages are basically all the same. There are gaps. We have many gaps in our language that other languages don't have, and conversely, they have gaps that we don't have. It's a little bit like what I said before about whether organisms vary infinitely or whether there's a universal genome. If you take a look at organisms, they look wildly different, so it was quite natural to assume fifty years ago that they vary in every possible way. The more we have learned, the less plausible that seems. There's a lot of conservation of genes. Yeasts have a genetic structure not all that different from ours in many ways, although yeasts look very different from us. But there are fundamental biological processes that just show up differently on the surface and seem different until you understand them. And something like that appears to be the case with language. Ken's work on this topic is the most sophisticated. There's a lot of popular discussion about "similar data" now, but most of it is extremely superficial and ignorant. In fact, there's almost nothing that's discussed now that he didn't talk about in a much more serious way forty years ago.

People who just read your books don't realize, I think, that you have a mischievous side. At the linguistics seminar I attended, I told you that I had to leave early, and you told me to shake my head back and forth, as I was leaving the classroom, and say, "I don't know what that guy Chomsky is talking about. This is just a lot of nonsense."

That's what this all sounds like if you don't have the right background. There's this commonsense idea: when I talk, I don't think about any of those things linguists are talking about. I don't have any of these structures in my head. So how can they be real? This kind of deep anti-intellectualism, an insistence on ignorance, runs through a large part of the culture. With discussions of language, it's almost ubiquitous.

You could say the same thing about vision. So, for example, one of the most interesting things known about the visual system is that it has core properties that interpret complex reality in terms of rigid objects in motion. In fact, you almost never see rigid objects in motion. It's not part of experience. But that's the way the visual system works.

Take, say, a baseball game. When you interpret an outfielder catching a fly ball, you don't and he doesn't introspect into the method by which he's doing that, which is a pretty remarkable thing. Like how does an outfielder know instantaneously where to run as soon as the crack of a bat takes place? It turns out that's a pretty sophisticated calculation and pretty well understood. But you can't introspect into it. In fact, if you did, you would fall on your face and you wouldn't catch the ball. It's sort of like trying to introspect on how you digest your food. You can't do it. People feel that they ought to be able to do it in cognitive domains because we're partially conscious—at least, we have a consciousness of some of the superficial aspects of our actions. For example, you know you're running to catch a ball. But consciousness of superficial aspects of our activity doesn't give you any insight into the internal computations of the brain that allow these actions to take place.

You've said many times that your linguistics and political work don't intersect in any way. But what is striking is your syncretic power, your ability to gather very disparate information together into a coherent picture.

I think anybody can do it. I have no special talents in that regard. There are some talents, if you like, that are useful for the sciences—or for the study of, say, international affairs or personal relations. One important one that everybody has, if they feel like using it, is just the ability to be puzzled. Why do things happen this way? If you look at the history of modern science, that ability has yielded dramatic results at many points. Albert Einstein was interested in the question of what the world would look like if you were traveling at the speed of light. He was puzzled by that. Out of that came important insights.

Modern science really developed from a willingness to question things that had always been taken for granted. If I have a cup in my hand full of boiling water, and I let go with both hands, the steam rises and the cup falls. Why? Well, for millennia there was a good answer from the best scientists: the cup and steam are both going to their natural place. The natural place of the steam is up there, and the natural place for the cup is down there. End of discussion. But Galileo and others decided that they were puzzled by this event. Why does this happen? And as soon as they started to be puzzled, the question turned out to be significant. As soon as you look carefully, you find that all your intuitions are wrong. Our intuition is that a heavy ball and a light ball will fall at different rates. They don't. In fact, just about all intuitions are wrong. Modern science comes out of that understanding.

When you go to the social and political domain, there are certain doctrines that are just taken for granted, like things go to their natural place. For example, the United States is a benign actor. It makes mistakes, but its leaders are trying to do good in the world. People make mistakes, it's a complicated world, but we're promoting democracy.

We love democracy. If you don't accept these dogmas, you're just not part of the discourse. That's true of ordinary discourse. That's true of professional scholarship to a remarkable degree. That's true of the media overwhelmingly. You can find case after case.

Take a look at an article in the *New York Times* by Bill Keller, the paper's former executive editor, on our inherent benign character.[2] He points out there are very troubling exceptions: we supported and are supporting serious atrocities in Bahrain, and we don't do anything about the most reactionary state in the region, Saudi Arabia. He says these exceptions are troubling because they don't fit our general nature. That's about at the level of "things go to their natural place."

It doesn't take much brilliance to recognize that this is not schizophrenia and there's nothing surprising about it. It's exactly the way great powers operate. They have domestic power structures that determine policy. There are a lot of other factors, but they're not overwhelmingly significant. If you look at the goals and intentions of political elites, everything falls into place. Of course, if you take that stand, you're excluded from polite discourse—just as, incidentally, Galileo was. He couldn't convince the funders, the aristocrats, that any of his ideas made any sense because they were so counter to common sense. He suffered for it under the Inquisition, as dissidents commonly do suffer. He was forced to renounce formally everything he believed. Legend has it that under his breath he said, "Eppur si muove" ("And yet it moves"). Whether that's true or not, I don't know.

Almost every society I've ever heard of, back to the earliest records, treats those we call dissidents, people who depart from the established consensus, pretty harshly. How harshly depends on the society. Another interesting thing about our culture is that we are very outraged by the harsh treatment of dissidents in enemy states. So the treatment of, say, Václav Havel or Aleksandr Solzhenitsyn is considered an utter outrage quite rightly. You can find countless articles in the *New York Times* about the horrible way in which dissidents are treated elsewhere. But, if you look at the facts, dissidents in U.S. domains are treated far more harshly. So you can read in the standard *Cambridge History of the Cold War* that since 1960 the record of torture, assassination, and other atrocities in U.S. domains vastly exceeds anything in the Soviet and Russian domains.[3] It's obviously true. So yes, Havel was imprisoned. Very bad. Six Jesuit intellectuals in El Salvador had their heads blown off.[4] Worse. In fact, nobody even knows their names. Everyone knows the names of the Eastern European dissidents. Try to find somebody who knows the names of the dissidents in, say, El Salvador or Colombia, anywhere in U.S. domains.

A lot of what is called the new media, Facebook and Twitter, plus what are called handheld devices, iPads and tablets and the like, are creating greater social atomization and isolation.

I've had the experience of being in a restaurant and everyone is looking down at their iPhone, sending messages and checking e-mail. What impact might this have on society?

I'm really not part of this culture at all, so I'm just observing it from outside, and not with very much intensity or understanding. But my impression is that the people participating in it, the young people participating in it have a feeling of intimacy and interaction. But I have to say, it reminds me of a close friend of mine as a kid who had a little booklet in which he wrote the names of all his friends. He used to boast that he had two hundred friends, which meant he had no friends, because you don't have two hundred friends. And I suspect that it's similar to that. If you have a whole bunch of friends on Facebook or whatever, it almost has to be pretty superficial. If that's your outlet to the world, there's something really missing in your life.

In fact, one of the significant aspects of the Occupy movements, maybe their most significant aspect, is the way they're overcoming that by creating real communities of people who interact, who have associations and bonds and help each other, support each other, really talk to each other freely, something which is very much missing in the whole society. You have it in bits and pieces, of course. But there has been, I think, a conscious effort to atomize the society for a long time, to break people up, to break down what are called secondary associations in the sociological literature: groups that interact and construct spaces in which people can formulate ideas, test them, begin to understand human relations and learn what it means to cooperate with each other. Unions were one of the major examples of this, and that's part of the reason for their generally very progressive impact on society. And, of course, they've been a major target of attack, I think partially for that reason.

The whole concept of social solidarity is considered very threatening by concentrated power. That's true in any system, and is very striking in ours.

Although the social media are undoubtedly invaluable for organizing and keeping some connections alive, I think they contribute to atomization. That's my superficial impression from outside.

Let's talk about education in a capitalist society. You've taught for many years. One of your strongest influences was the educator John Dewey, whom you've described as "one of the relics of the Enlightenment classical liberal tradition."[5]

One of the real achievements of the United States is that it pioneered mass public education, not just elite education for the few and maybe some vocational training, if anything, for the many. The opening of land-grant colleges and general schools in the nineteenth century was a very significant development. But if you look back, the reasons for this were complex. Actually, one of them was discussed by Ralph Waldo Emerson. He was struck by the fact that business elites—he didn't use that term—were interested in public education. He speculated that the reason was that "you must educate them to keep them from our throats."[6] In other words, the mass of the population is getting more rights, and unless they're properly educated, they may come after us.

There's a corollary to this. If you have a free education that engenders creativity and independence, the way of looking at the world that we were talking about before, people are going to come for your throat because they won't want to be governed. So yes, let's have a mass education system, but of a particular kind, one that inculcates obedience, subordination, acceptance of authority, acceptance of doctrine. One that doesn't raise too many questions. Deweyite education was quite counter to this. It was libertarian education.

The conflicts about what education ought to be go right back through the early Enlightenment. There are two striking images that I think capture the essence of the conflict. One view is that education should be like pouring water into a bucket. As we all know from our own experiences, the brain is a pretty leaky bucket, so you can study for an exam on some topic in a course you're not interested in, learn enough to pass the exam, and a week later you've forgotten what the course was. The water has leaked out. But this approach to education does train you to be obedient and follow orders, even meaningless orders. The other type of education was described by one of the great founders of the modern higher education system, Wilhelm von Humboldt, a leading figure and founder of classical liberalism. He said education should be like laying out a string that the student follows in his own way.[7] In other words, giving a general structure in which the learner—whether it's a child or an adult—will explore the world in their own creative, individual, independent fashion. Developing, not only acquiring knowledge. Learning how to learn.

That's the model you do find in a good scientific university. So if you're at MIT, a physics course is not a matter of pouring water into a bucket. This was described nicely by one of the great modern physicists, Victor Weisskopf, who died some years ago. When students would ask him what his course would cover, he would say, "It doesn't matter what we cover. It matters what you discover." In other words, if you can learn how to discover, then it doesn't matter what the subject matter is. You will use that talent elsewhere. That's essentially Humboldt's conception of education.

I should say that I learned about this not from books but from experience. I was in a Deweyite experimental school. That was the way things worked. It seemed very natural. I only read about it later.

The battle over education has been going on for quite some time now. The 1960s were a major period of agitation, activism, exploration, and they had a major civilizing effect on the society: civil rights, women's rights, a whole range of things. But for elites, it was a dangerous time because it had too much of a civilizing effect on the society. People were questioning authority, wanting to know answers, not just accepting everything that was handed down. There was an "excess of democracy."[8]

Looking for answers—that's frightening. There was an immediate backlash in the 1970s, and we're still living with the results. All of this is well documented. Two of the striking documents, which I think are very much worth reading, from opposite ends of the spectrum, are, on the Right, the Powell memorandum and, on what's called the Left, the Trilateral Commission report.

Lewis Powell was a corporate lobbyist for the tobacco industry who was very close to Nixon, who later appointed him to the Supreme Court. In 1971, he wrote a memorandum to the Chamber of Commerce, the main business lobby.[9] It was supposed to be secret but it leaked. It's quite interesting reading, not only for the content but also because of the style, which is pretty typical of business literature and of totalitarian culture in general. It reads a little like NSC-68.[10] The whole society is crumbling, everything is being lost. The universities are being taken over by followers of Herbert Marcuse. The media and the government have been taken over by the Left. Ralph Nader is destroying the private economy, and so on. Businessmen are the most persecuted element in the society, but we don't have to accept it, Powell said. We don't have to let these crazy people destroy everything. We have the wealth. We're the trustees of the universities. We're the people who own the media. We don't have to let all this happen. We can get together and use our power to force things in the direction that we want—of course he used nice terms such as democracy and freedom.

It is such a grotesque caricature, you have to wonder what lunacy could allow people to think like this. But it's normal. Like a three-year-old who doesn't get his way, if you think you ought to own everything and you've lost anything, everything is gone. That's very much the attitude of those who are accustomed to power and believe they have a right to power.

At the other end of the spectrum, you have the Trilateral Commission report, *The Crisis of Democracy*, written by liberal internationalists, Carter administration liberals, basically. They were concerned about what they called the failure of the "institutions which have played the major role in the indoctrination of the young."[11] The young are not being properly indoctrinated by the schools, the churches. We can see that from the pressures for too much democracy. And we have to do something about it. It's not very different from Powell's memorandum. It's a little more nuanced, but it's essentially the same idea.

Too much freedom, too much democracy, not enough indoctrination—how do you deal with that? In the educational system, you move toward more control, more indoctrination, cutting back on the dangerous experiments with freedom and independence. That's what we've seen. These shifts correspond to the period when corporatization of the universities began to take place, with a sharp rise in managerial structures and a "bottom line" approach to education, and also when tuitions start to rise. The tuition problem has become so huge that it's on the front pages now. Student debt is on the scale of credit-card debt and by now it probably exceeds it.[12] Students are burdened by huge debts. The laws have been changed so there's no way out—no bankruptcy, no escape.[13] So you're trapped for life. That's quite a technique of indoctrination and control.

There's no economic basis for rising tuition costs. In the 1950s, our society was much poorer, but education was essentially free. The GI bill, was, of course, selective—it was for whites, not blacks, mostly men, not women—but it did offer free education to a huge part of the population that never would have gotten to college otherwise.[14] More broadly, tuition was very low by current standards. It was a great help to the economy, incidentally. The 1950s and the 1960s were the decades of the greatest economic growth in history, and the newly educated population was a significant part of that story.

Now we're a much richer society than we were in the 1950s. Productivity has increased a lot. There's way more wealth. So it's ludicrous to think that education can't be funded. The same conclusion can be drawn by looking at other countries. Take, say, Mexico. It's a poor country. It has quite a decent higher education system. The quality is high. Teacher salaries are low by our standards, but the system is quite respectable. And it's free. Actually, the government did try some years ago to add a small tuition, but there was a national student strike and the government backed down.[15] So education in this poor country is still free. The same is true in rich countries such as Germany and Finland, which has the best education system in the world by many measures.[16] Education in these countries is free—or virtually free. If you look at the percentage of our gross domestic product that would be required to provide free higher education, it's very slight. So it's very hard to argue that there are any fundamental economic reasons for rising tuition costs. But it does have the effect of control and indoctrination.

Look at K-to-12 education, kindergarten through high school. Policies like No Child Left Behind under Bush and Race to the Top under Obama, despite what they may claim, basically require schools to teach to the test. They control teachers and make sure that they don't move in independent directions, a step toward imposing a business model, as in the colleges. Anyone who has any experience with the K-to-12 system knows how this works. Students are required to conform, to memorize to pass the next test. And there are punitive measures to keep teachers in line. If the students don't get a high-enough grade on the test—which could mean they're too creative and independent—then the teacher is in trouble. So they are forced to conform to this system.

Meanwhile, the basic problems with the educational system are never addressed. It's just way underfunded. Class sizes are too large. Diane Ravitch, formerly a conservative education critic who is now very critical of the current system and very knowledgeable, recently did some comparative work on the Finnish educational system, which gets all the best records in the world. She showed that one of the major differences is that teachers are respected in Finland.[17] Teaching is considered a respected profession. Good people go into the field. They put energy and initiative into their work. They're given a good deal of freedom to experiment, explore, let students search on their own.

In *Science*, the journal of the American Association for the Advancement of Science, Bruce Alberts, a biochemist, had a series of editorials on science education.[18] What he points out is quite interesting. He says science education is increasingly being designed with the effect of killing any interest in science. If you are in college, maybe you have to memorize a bunch of enzymes or something. If you are in elementary school, you memorize the periodic table. When you study the discovery of DNA, you're just taught what scientists already discovered. You memorize the fact that DNA is a double helix. Science is being taught in a way that kills any joy in science, gives you no sense of what discovery is. It's the opposite of Weisskopf's view that it matters what you discover, not what you cover.

Alberts gives some nice examples of alternatives that

do work. In one kindergarten class, each kid was given a dish with a mixture of pebbles, shells, and seeds, and asked, "How do we know if something is a seed?"[19] So the class began with what they called a "scientific conference." The kids got together and discussed the various ways in which you might be able to figure out what a seed is. The kids were guided by the teacher, so if things went off in some wrong direction, the teacher could step in. But it's essentially laying down the string. Here's your task. Figure it out. Over time, they did figure it out. They ran some experiments, tried out new ideas, interacted. At the end of this particular program each kid was given a magnifying glass. They cut open the seeds and discovered what the embryo is that gives the seed its energy and differentiates it from a pebble. Those kids learned something. Not only did they learn something about seeds, which doesn't matter that much, they learned what it is to discover something, why it's fun and exciting, why you should try it somewhere else, why you should be puzzled and inquire.

That can be done at any level of education. A friend of mine who teaches sixth grade described to me once how she had taught her students about the American Revolution. A couple of weeks before they got to that assignment, she started acting very harshly, issuing orders and commands, making the kids to do all kinds of things they didn't want to do. They got pretty upset, and they wanted to do something about it. They started to get together and protest. By the time it got to the right point, she opened the lesson on the American Revolution. She said, "Okay, now you can see why people rebel." And they understood why you would. That's the type of creative teaching that doesn't pass some standardized test necessarily, but it allows children to learn. That can be done at any level, from kindergarten to graduate school, in any subject—history, science, whatever it may be.

So those are the two concepts. And it's pretty clear which way the educational system is being pressed—and I think there's a reason why. We've got to educate people to keep them from our throats, as Emerson put it a long time ago. At the K-to-12 level, there is now an effort to destroy the public educational system. That's essentially what charter schools are about. They don't have any better outcomes. They feed at the public trough, the public pays for them, but they're essentially out of the public system and under much more private control, essentially privatized. It's destroying the ethic of the public education system. The ethic of that system is solidarity. You have a public education system because you're supposed to care whether children you don't know and have nothing to do with have the opportunity to go to school. That's social solidarity, but that's very dangerous—the opposite of atomization.

My feeling is that Social Security is under attack for the same reasons. There's no economic reason. It's in very good shape. With a little tinkering, it could go on indefinitely.[20] But it's always listed as one of the big problems. We've got to do something about Social Security. I think the issue is the same: it's a system based on the concept that you should care about others, that you should care whether elderly people you don't know can live decent lives. You can't have that sort of thing. If a widow somewhere doesn't have food, it's her problem. She married the wrong husband or didn't invest properly. In a society in which everyone is out just for themselves, you don't pay attention to anyone else.

Ron Paul was asked at a Republican presidential debate what if "something terrible happens" to some guy who has no health insurance? What do you do? He said, "That's what freedom is all about: taking your own risks."[21] Actually, when the moderator pushed back on this, he backed off and he said that people without health insurance would be taken care of by their families or their church. Then Rand Paul—this is more interesting—said national health insurance is slavery.[22] He said, I'm a physician, and if there's national health insurance, the government is forcing me to take care of somebody who is ill. Why should I be a slave to the state? Here we're getting capitalist pathology in its most extreme, lunatic form. It is the opposite of solidarity, mutual support, mutual help.

Is it a form of social Darwinism?

I wouldn't even call it social Darwinism. That's too sophisticated. It's just, I'm out for myself, nobody else—and that's the way it ought to be. There was a recent study done at Harvard University's Institute of Politics on attitudes of young people from ages eighteen to twenty-nine.[23] It was pretty striking. There's a lot of commitment to what in the United States are called libertarian ideas. Libertarian in the United States is pretty close to totalitarian. If you really think through what are called libertarian concepts, they basically say that we're going to hand over decision making to concentrations of private power and then everybody will be free. I'm not saying the people who advocate it intend that, but if you think it through, that's the consequence, plus the breaking down of social bonds. A lot of young people are attracted to that. For example, less than half of the people in the Harvard survey felt that the government should provide health insurance or "basic necessities, such as food and shelter" to those in need who cannot afford them.[24]

When people talk about the government in the United States, they're talking about some alien force. Hatred of democracy is so deeply embedded in the doctrinal system that you don't think of the government as your instrument. It's some alien instrument. It's taken a lot of work to make people hate democracy that much. In a democratic society, to the extent that it's a democratic society, the government is you. It's your decisions. But the government here is depicted as something that's attacking us, not our instrument to do what we decide.

Actually, one of the most frightening statistics for the Harvard survey has to do with the environment. Only 28 percent think that the "government should do more to curb climate change, even at the expense of economic growth."[25] If that continues, that's a death sentence for the species. But it's the anticipated result of the major attack on social solidarity, on participation, on interaction, and on the fundamentals of democracy.

April 15, the day when you pay your taxes, gives you a good index of how democracy is functioning. If democracy were functioning effectively, April 15 would be a day of celebration. That's a day on which we get together to contribute to implementing the policies that we've decided on. That's what April 15 ought to be. Here it's a day of mourning. This alien force is coming to steal your hard-earned money from you. That indicates an extreme contempt for democracy. And it's natural that a business-run society and doctrinal system should try to inculcate that belief.

8

ARISTOCRATS AND DEMOCRATS

CAMBRIDGE, MASSACHUSETTS (MAY 15, 2012)

There was a big sex scandal around the Cartagena Summit of the Americas in Colombia in spring 2012, but in a column for the New York Times Syndicate *you pinpointed some more substantive developments.*[1]

It was actually a very interesting and significant conference. The participants didn't come out with a formal declaration because they couldn't reach agreement. The reason they couldn't reach agreement was that, on the two major issues, the United States and Canada rejected what the rest of the hemisphere insisted on, which was inclusion of Cuba and serious consideration of the decriminalization of drug policy.[2] That's very significant, another step in the isolation of the United States and

Canada—and in the integration of the Latin American and Caribbean countries, which is very important.

About a year ago, a new organization formed called CELAC, the Community of Latin American and Caribbean States.[3] CELAC includes all the countries in the hemisphere minus the United States and Canada. There is some belief that it might actually replace the traditionally U.S.-dominated Organization of American States. There are already steps in that direction, with UNASUR, the Union of South American Nations, which has functioned with some success in a number of cases.

Latin America also has shown increasing independence in international affairs. Brazil, for example, has taken on a very interesting role in the international system, which the United States doesn't like.

If there is another hemispheric summit and Cuba is admitted, the United States will presumably stay home. Or if the United States blocks Cuba's participation again, there just won't be a summit. Washington is also isolated on its position on drugs. More and more countries in the hemisphere are moving to change drug policy. Even conservative presidents are calling for decriminalization. Not legalization, but shifting possession of drugs from a criminal offense to an administrative matter, like a parking ticket. These policies have been pretty successful in Europe. That's essentially what most of Latin America is moving toward, beginning with marijuana, maybe moving on to other drugs. Again, the United States just refuses flat out.

It's quite significant, because the people of Latin America and the Caribbean are the victims of these policies. In Mexico alone, tens of thousands of people have been killed in the drug-related violence. And the United States is the source of the problem, a dual source, actually—in terms of demand, which is obvious, and also supply, which is hardly discussed. The guns to the Mexican cartels are increasingly from the United States. The Bureau of Alcohol, Tobacco and Firearms, a federal government bureau, analyzed guns that were confiscated in Mexico. According to their figures, about 70 percent came from the United States.[4] Furthermore, the type of guns has been shifting over the years. A couple of years ago, maybe people were smuggling in pistols, now it's assault rifles.[5] Next year who knows what it will be?

This is all connected to the crazy gun culture in the United States. I don't know if you saw this, but Rand Paul just came out with an appeal for a new organization that will counter the efforts by Obama and Hillary Clinton to shred the last remnants of our sovereignty by allowing the United Nations to take away our guns.[6] And then, of course, they will come and conquer us. The basis for this is that the UN is now debating a small arms treaty.[7] Small arms doesn't mean pistols. It means anything less than a tank. These are just slaughtering people all over the world. Hundreds of thousands of people every year are killed with small arms, and a high percentage come from the United States.[8] So there is an effort to have some sort of small arms treaty to regulate their flow. In the minds of the Rand Paul libertarians, this is just another effort by this ominous, fiendish outfit, the United Nations, to take away our freedom.

Rand Paul is the Republican senator from Kentucky and son of Ron Paul.

And apparently is being groomed as the future of libertarianism or something.

What about Canada's role in all of this? Why is Ottawa so yoked to Washington's policy?

It's an interesting development in recent years. It's related to NAFTA, but it reflects more general trends. Canadian and U.S. capital are increasingly integrated, which is bringing elites closer together. You can ask about cause and effect but Canadian policies, particularly under Stephen Harper, the prime minister, are not just drawing closer to U.S. policies but in some cases even going beyond them in extremism. Canada is becoming less and less of an independent country in many respects, culturally, economically, politically. It's increasingly embedded within the U.S.-run system as a kind of client state.

The energy system is a key part of this integration. The

tar sands in Canada are a huge source of potential energy—and of environmental destruction. There's a controversy going on about who is going to exploit the tar sands. The United States wants to, but Canada occasionally warns that it will partner with China, which is eager to develop these fields if the United States won't.[9] This is a major issue now. In his 2012 State of the Union address, Obama was very enthusiastic about the idea that we could have a century of energy independence by making use of fossil fuels in North America—natural gas in the United States and fuel from tar sands.[10] He didn't talk about what kind of a world it would be in one hundred years if we use these fossil fuels. There's some discussion of the local environmental effects of developing the Canadian tar sands, but there's a much broader question about the general effect on the global environment. These are very serious issues.

Canada is also one of the major centers of mining operations around the world. Conflicts over mining of natural resources are leading to wars and violence globally, from Latin America to India. Internally, India is practically at war over natural resources.[11] The same is true of Colombia and other countries.

What can you say about the process of hydraulic fracturing to extract natural gas, known as fracking?

Fracking has local environmental ramifications that are pretty severe. It uses huge amounts of water. The process itself is destructive of the local environment in many respects, and there is considerable public opposition to it on that basis.[12] But I think that we shouldn't overlook the deeper problem. Suppose it were environmentally pure. You're still using fossil fuels. And we are coming to a tipping point on fossil fuels. We can't continue in this direction for long without getting to a point of irreversible devastation. You can't be sure of the date, but it's pretty clear that it's coming.

The Vikings football team was threatening to move to Los Angeles, so the good taxpayers of the state of Minnesota will provide almost half a billion dollars in public money for the construction of a new stadium to keep the team there.[13]

Florida also announced recently that it's cutting back funding for the state university. The University of Florida is getting rid of some major academic programs, including computer science, but increasing funding for sports.[14]

Athletic departments on U.S. campuses operate in a separate world. The salaries of the coaches are in the millions of dollars.[15]

I remember going to some college for a talk—I forget where it was—but the first thing we drove to was some huge stadium. Right next to the sports stadium was a big building. I asked the students what it was, and they said,

"That's where the football players live." They get special training to enable them to pass the courses so they can keep playing football.

Years ago, you talked about listening to talk radio sports shows. I don't know if you're still doing that.

I still do.

I remember at the time you commented that these talk shows give the lie to the idea that the average Joe is unable to grasp complex and arcane data. And also, callers demonstrate fearlessness. You hear them saying, "Fire the bum," "Get rid of that coach," "Trade that player."

It's very striking. First of all, there's an enormous amount of knowledge, and a lot of self-confidence and challenging of authority, which is normal. If you don't like what the coach did, you say he made a stupid decision, get rid of him. We're smarter than he is. If you could carry that over to other domains of life, it would have some significance.

I don't know if you've read that your hometown, Philadelphia, is closing forty of its public schools.[16]

I didn't see that, but it's happening elsewhere, too. I was invited a couple of months ago by a black community in Harlem to give a talk at one of the churches there, a famous church with a long civil rights history. They wanted me to talk about education. And a lot of the concerns people articulated there were that the public education system is under serious attack, both by defunding and by charter schools, which are breaking up the community and undermining the basic contributions of the public education system, which are quite real in the black community.

In California, which is one of the richest places in the world, but is now under severe budgetary constraints, the major public universities, Berkeley and UCLA, the jewels in the crown, are effectively being privatized. They're not very different from Ivy League universities now. Tuitions are sky high. They have endowments. At the same time, the state college system is being downgraded, so much so that students and teachers are planning a rolling strike against the budgetary cuts.[17] California State University announced that it's just going to have to refuse to accept any students for the spring 2013 term.[18] The educational system is being degraded for the general population. But you have a private education for the rich and the privileged, and some small group that will be selected out to receive scholarships. It's a sharply two-tiered system.

One of the amazing things that's happened in recent years is the corporatization of the universities, which shows up in many ways. There's been a rapid increase in the number of administrators and layers of administration. They bring in a corporate mentality. Each new administrator has to have a sub-administrator, and that one has to have a sub–something else. Meanwhile, the role of the faculty in running the university is sharply declining. There's a useful book on this topic by Benjamin Ginsberg called *The Fall of the Faculty*.[19]

All of these developments are part of the general assault on education, which we should remember is part of a much more general assault on the whole society. That's the neoliberal program, which is being protested all over the world, by the Occupy movement here, by the activists in Tahrir Square in Egypt, in different forms in different countries, but all over. It's a very harmful system, except for the very rich. Actually, there's a nice little monograph that just came out from the Economic Policy Institute—which is the main source of reliable, regular data on working America and the economy—called *Failure by Design*.[20] The author, Josh Bivens, reviews the economic policies of the past forty years roughly and points out that they're a class-based failure. Of course, they're a great success for the top tenth of a percent of the population—the traders, CEOs—but they're a failure for the large majority. By design. There are plenty of alternative policies, but these others are the ones that are chosen.

We're seeing similar dynamics right now in dramatic form in Europe, where the banks and the bureaucrats have been imposing a policy of austerity under stagnation, which is almost bound to make things worse and will make it harder to pay debts. They've been pretty sharply criticized by economists, even by the business press, but they're pursuing austerity. It's difficult to give a rationale on economic grounds. In fact, I think impossible. But you can find a rationale. In fact, it was more or less stated by the president of the European Central Bank, Mario Draghi, in an interview in the *Wall Street Journal* in which he said that the social contract in Europe is over.[21] In other words, we're killing the social contract.

You always talk about the institutional imperatives and the structural underpinning of these policies. But don't you have to keep the patient in reasonably good health and functioning? Aren't they killing the goose?

It depends on what time scale you have in mind. There's plenty of cheap labor around the world. You can outsource production. If you're Apple, one of the world's richest corporations, you can have workers employed by Foxconn, a Taiwanese company, in southwestern China, living and working in hideous conditions, committing suicide, and you can make a lot of profit out of that.[22] If China turns out to be too expensive, you can go to Bangladesh and sub-Saharan Africa. You can keep that going for a long time. Yes, there's a long-term problem, but there are long-term problems in capitalist economies anyway. There's a problem with overproduction. There's a crisis of accumulation. These are long-term problems that you try to keep at bay in various ways, all while planning for short-term wealth and privilege. That's the way business works.

Apple also, conveniently, has an office in Reno, Nevada, and by doing so, according to a recent report, "it has avoided millions of dollars in taxes in California" and other states.[23] *Meanwhile, California is cutting programs left and right.*

It's a standard technique. It's now called globalization. It's been going on for quite a while.

Robert Reich was secretary of labor during the Clinton administration. He's currently a media pundit and professor at Berkeley. He says, in France, "Socialism isn't the answer to the basic problem haunting all rich nations. The answer," he says, "is to reform capitalism. . . . We don't need socialism. We need a capitalism that works for the vast majority."[24] *What do you think about this idea of sustainable capitalism?*

In a narrow sense, I agree with him. If we're talking about feasible objectives in the short term, it's kind of meaningless to talk about socialism. There isn't a popular base for it. There isn't an understanding of it. That's, of course, not what he means—but if we keep to that narrow frame, yes, there's a point.

In the long term, it's almost a self-contradiction. Capitalism is based on production for profit, not need. It's also based on a requirement of constant growth for profit. That's self-destructive—quite apart from things like the steady process of monopolization, forming more and more oligopolies, as well as overproduction and the decline of the rate of profit. These are long-term tendencies that can be delayed, but they're inherent to capitalism.

And, at least from my point of view, there's something essentially wrong with the current system. Here we get to values. Do we want to have a system in which some people give orders and others take them? That's a deeper question. Do we want it in the political system? Do we want it in the economic system, especially given the inevitable interaction between the two, with concentration of wealth heavily influencing political power? Or should we be moving toward enterprises that are owned and managed by the workforce and communities? Call it whatever name you want. You can call it capitalist if you like. You can call it anything. But that's a direction in which policy could move: toward more democracy, undermining illegitimate authority.

Various moves in this direction are taking place. There's a new organization being formed, an International Organization for a Participatory Society, coming largely out of the ZNet collective.[25] The United Steelworkers has a new initiative with Mondragon, the huge worker- and community-owned conglomerate in the Basque country in Spain.[26] Mondragon runs industrial enterprises, banks, schools, hospitals, housing. It's quite successful economically and quite complex. Mondragon functions in an international capitalist economy—a quasi-market economy—which often has ugly consequences. But things like Mondragon could still become what Mikhail Bakunin once called "the seeds of the future" in the present society.[27] I don't know what Robert Reich thinks about this, but I would think that's a much saner way to move in the long term.

In a lecture at Loyola University, you pointed out that Thomas Jefferson had rather serious concerns about the fate of the democratic experiment.[28] He feared the rise of a new form of absolutism that was more ominous than the British rule overthrown in the American Revolution. He distinguished in his later years between what he called "aristocrats and democrats."[29] And then he went on to say, "I hope we shall . . . crush in its birth the aristocracy of our moneyed corporations, which dare already to challenge our government to a trial and bid defiance to the laws of our country."[30] He also wrote, "I sincerely believe . . . that banking establishments are more dangerous than standing armies."[31] That's the kind of quote from a Founding Father you don't see too much.

Yes, those lines are not usually quoted. But these concerns were felt early on, for a complex variety of reasons. They take newer forms constantly.

Didn't Bakunin say, "If there is a State, there must be domination of one class by another"?[32]

He did, but I would somewhat take issue. The state is not the only center of power in our society. In fact, there's another center of power: concentrated private capital. And as long as that's there, in many ways the state is a protector against its excesses. So I think he's right in criticizing the state as an oppressive institution, but it is also relative to the nature of the rest of the society.

Bakunin was not a systematic thinker, but he did have significant insight into the nature of power and its exercise. His disagreements with Karl Marx were over an element of that conflict. He objected to what he understood to be Marx's conception of a kind of a radical intelligentsia running the workers' movement, for its own benefit, of course. He pointed out, very presciently, that what he called a new class of scientific intelligentsia, who claim to appropriate all knowledge, would move in one of two directions: either it would become a "red bureaucracy," which would institute the most oppressive rule ever seen in the name of the working classes, or it would recognize that power lies elsewhere, in private capital, and become its servants.[33] That's essentially what happened. That's a pretty good prediction—one of the few far-reaching predictions in the social sciences that really came true. It should be studied everywhere for that reason alone.

There's a movement in the country to reverse the Citizens United v. Federal Election Commission *ruling of the Supreme Court in January 2010, which deregulated the campaign finance system, and which, one critic says, "has legalized corporate bribery of our elected officials."[34] What are your views on* Citizens United *and the efficacy of engaging in such a constitutional amendment ratification, which may take years and years?*

There are a number of questions, including a tactical question of the kind you raise and a principled question, the heart of the matter. And there's something to say about each of them. On the tactical side, I think a campaign to amend the Constitution could be justified as an educational effort, a way to get people to pay attention to the issue. That's independent of how long it might take to ratify something like that. If enough people get interested in the issue, they may turn to more radical goals and, I think, more principled ones. Which takes us to the principled issue. I think *Citizens United* is a very bad decision. However, it's kind of the icing on the cake. The idea of corporate personhood goes back a century. It wasn't instituted by *Citizens United*. And we should be thinking about that.

Why should corporations be granted personal rights? By now corporations have rights way beyond persons of flesh and blood. They are immortal, they are protected by state power. In fact, the basis of a corporation is limited liability, meaning as a participant in a corporation you're not personally liable if it, say, murders tens of thousands of people at Bhopal.

You are referring to the Union Carbide explosion in Bhopal, India, which killed about twenty thousand people in 1984.[35]

Which is just one example. Why should such an institution have personal rights? Also, these institutions are directed to maximize shareholder rights at the expense of stakeholder rights by law. Why should we accept that? It's not an economic principle, certainly.

Under NAFTA, U.S. corporations have the right of what's called "national treatment" in Mexico.[36] A Mexican person doesn't have the right of "national treatment" in Arizona, obviously. Why should a corporation have such rights?

Another major Supreme Court decision, *Buckley v. Valeo,* back in the 1970s, interpreted money as a form of speech.[37] That has far-reaching implications. If money is a form of speech, then those who have money can shout louder. I should say, the American Civil Liberties Union has supported these judgments on the basis of a form of free-speech absolutism.[38] I don't think they're thinking through the implications.

Citizens United opens the way for massive contributions that distort the political system.[39] But this is something that's been going on for a long time. So we're talking about an expansion of something that shouldn't have happened in the first place.

Marx said that the task is not just to understand the world but to change it.[40] *You've devoted much of your life to that.*

For whatever it's worth—that's for others to decide. But sure, I think that's what we should all be trying to do: change the world in the short term, overcoming immediate problems—some of them, like environmental disaster and nuclear war, lethal problems. Not small problems. The fate of the species depends on them. So, in the short term, you can work for what are called reforms. Others try to get at the heart of the forms of illegitimate authority, dismantle them, and move toward greater freedom and independence.

But you warn that victories don't come quickly. So we're not in a sprint here. It's a marathon.

It's a marathon, and one in which you often go backward. There's regression, too. The last thirty years have been in some respects a period of regression, although in popular activism it's been an expansion. History is never simple.

You've got grandchildren. What kind of world do you see them inheriting?

A realistic projection would not be very attractive. But a lot depends on human will, as always. You cannot predict the course of social movements, of the efforts to change things. We can never do that. Nobody could have predicted in 1960 that a handful of black students sitting in at a lunch counter in Greensboro, North Carolina, would help set off a massive civil rights movement. And nobody could have predicted in the early stages of the women's movement that it would radically change the culture, as it has in a very effective way. If you had asked me a year ago, "Does it make sense to occupy Zuccotti Park?" I would have said you're out of your mind. That couldn't possibly work. It worked spectacularly well. How it goes on from here, it depends.

And what advice would you give to young people starting out?

Every night when I come home and start to answer the day's hundreds of e-mails, a fair number are from young people saying, "I don't like the way the world is going. In fact, I can't stand it. What should I do?" By now I receive so many that I'm almost compelled to resort to form responses. And what I point out is that you're well on the way to answering the question yourself, because you recognize there's a problem. There is no general answer for everybody. There is no right answer for every person, in all circumstances. It depends on who you are, what your concerns are, what your options are, how much you want to devote yourself to it, what your talents are. But you're probably pretty privileged. Otherwise you wouldn't be writing me a letter on the Internet. That means you have a lot of opportunities—much more than your counterparts in other countries, or even here a generation ago. So there is a legacy that you can use. It's not going to be easy—it never is. But you can make a difference. You just have to find your own way.

There's no way people can answer for anyone else the questions, "What should I devote my life to? How should I live?" Those are things you just have to work out for yourself. You will go down paths that don't work. There will be failures, which you can learn from and then go back and start over in a different direction. It's in your hands.

Forgive me for saying this, but you're in your mid-eighties. Are you going to keep up your punishing travel and speaking schedule? You've essentially retired from teaching, right?

Yes, although I do work with students and sometimes teach and lecture, of course. I'll try to keep up both sides of my life as long as I can do it. I have no profound things to say about it. I don't expect a long period ahead, but I'll do what I can.

But your health is holding up?

Reasonably well. No complaints.

NOTES

1. THE NEW AMERICAN IMPERIALISM

1. For details on Vietnam, see Noam Chomsky, *At War with Asia: Essays on Indochina* (Oakland, CA: AK Press, 2004). See also Noam Chomsky, *For Reasons of State* (New York: New Press, 2003) and *Rethinking Camelot: JFK, the Vietnam War, and US Political Culture* (Cambridge, MA: South End Press, 1999).
2. Noam Chomsky, *Year 501: The Conquest Continues* (Cambridge, MA: South End Press, 1993), p. 22.
3. Bernard Porter, *Empire and Superempire: Britain, America and the World* (New Haven, CT: Yale University Press, 2006), p. 64.
4. See Philip S. Foner, *The Spanish-Cuban-American War and the Birth of American Imperialism*, 2 vols. (New York: Monthly Review Press, 1972).
5. For further discussion, see Noam Chomsky, *Hopes and Prospects* (Chicago: Haymarket Books, 2010).
6. Simon Romero, "Ecuador's Leader Purges Military and Moves to Expel American Base," *New York Times*, 21 April 2008.
7. Hugh O'Shaughnessy, "US Builds Up Its Bases in Oil-Rich South America," *Independent* (London), 22 November 2009.
8. Staff, "Controversial Agreement," *Panama Star*, 29 September 2009.

I. Roberto Eisenmann Jr., "Be Careful That with the Drug Story, We Return to Having Bases Again!" *La Prensa*, 2 October 2009.

9. For discussion, see Mark Weisbrot, "More of the Same in Latin America," *New York Times*, 11 August 2009.
10. Charlie Savage, "DEA Squads Extend Reach of Drug War," *New York Times*, 7 November 2011. See also John Lindsay-Poland, "Beyond the Drug War: The Pentagon's Other Operations in Latin America," *NACLA Report on the Americas*, 26 August 2011.
11. For further discussion, see "Militarizing Latin America," Chomsky.info, 30 August 2009. Available at http://chomsky.info/articles/20090830.htm. See also William M. LeoGrande, "From the Red Menace to Radical Populism: U.S. Insecurity in Latin America," *World Policy Journal* 22, no. 4 (Winter 2005–06), pp. 25–35.
12. James Zacharia, "Obama Backing of Honduras Election Crimps Latin Ties," Bloomberg News, 27 November 2009.
13. James Gerstenzang and Juanita Darling, "Clinton Extols Mitch Relief Efforts by GIs," *Los Angeles Times*, 10 March 1999.
14. Kirsten Begg, "Colombia and Honduras Sign Anti-Drug Trafficking Pact," *Colombia Reports*, 15 February 2010. See also "Honduran, Colombian Presidents Sign Agreement," BBC Latin America, 24 May 2011.
15. Daniel Kruger, "Japan Overtakes China as Largest Holder of Treasuries," Bloomberg News, 16 February 2010.
16. For data, see the regular report of the Federal Reserve Board, Department of the Treasury, "Major Foreign Holders of Treasury Securities." Available at http://www.treasury.gov.
17. Adam Smith, *The Wealth of Nations: Books IV–V* (New York: Penguin Books, 1999), p. 247.
18. Ibid., p. 25.
19. Francisco Rodriguez and Arjun Jayadev, *The Declining Labor Share of Income*, United Nations Development Programme, Human Development Research Paper 2010/36 (November 2010). See also Eva Cheng, "China: Wage Share Plunges," *Green Left Weekly*, 19 October 2007.
20. For an analysis, see Paul Mason, *Live Working or Die Fighting: How the Working Class Went Global*, updated ed. (Chicago: Haymarket Books, 2010).
21. United Nations Development Programme, *Human Development Report 2011: Sustainability and Equity: A Better Future for All* (New York: United Nations Development Programme, 2011), p. 126.
22. See Arundhati Roy, "Beware the 'Gush-Up Gospel' Behind India's Billionaires," *Financial Times* (London), 13 January 2012.

23. Arundhati Roy, *Field Notes on Democracy: Listening to Grasshoppers* (Chicago: Haymarket Books, 2009), p. 55.
24. BBC, "Carlos Slim Overtakes Bill Gates in World Rich List," 11 March 2010.
25. Ching Kwan Lee, *Against the Law: Labor Protests in China's Rustbelt and Sunbelt* (Berkeley: University of California Press, 2007).
26. See Tom Mitchell, "China: Strike Force," *Financial Times* (London), 10 June 2010.
27. Smedley Butler, "America's Armed Forces: 'In Time of Peace,'" *Common Sense* 4, no. 11 (November 1935), p. 8. See also Smedley Butler, *War Is a Racket* (Los Angeles: Feral House, 2003).
28. For discussion, see Chomsky, *Year 501*, chap. 8.
29. Butler, *War Is a Racket*, pp. 11–12.
30. Alissa J. Rubin and Helene Cooper, "In Afghan Trip, Obama Presses Karzai on Graft," *New York Times*, 28 March 2010.
31. "What Obama Told U.S. Troops in Afghanistan," *Los Angeles Times*, 28 March 2010.
32. Walter Pincus, "Mueller Outlines Origin, Funding of Sept. 11 Plot," *Washington Post*, 6 June 2002.
33. Michael R. Gordon, "Allies Preparing for Long Fight as Taliban Dig In," *New York Times*, 28 October 2001. Boyce told the *Times*, "I believe that what we should try to do is not let them think that we are going to give up and go away or lighten up. . . . The squeeze will carry on until the people of the country themselves recognize that this is going to go on until they get the leadership changed."
34. Abdul Haq, "US Bombs Are Boosting the Taliban," *Guardian* (London), 2 November 2001. Excerpted from an 11 October 2001 interview with Anatol Lieven.
35. Farhan Bokhari and John Thornhill, "Afghan Peace Assembly Call," *Financial Times* (London), 26 October 2001.
36. Mehreen Khan, "Iran Builds New Gas Pipeline," *Financial Times* (London), 6 July 2011.
37. Peter Baker, "Senate Approves Indian Nuclear Deal," *New York Times*, 1 October 2008.
38. For poll data, see "Pakistani Public Turns Against Taliban, But Still Negative on US," World Public Opinion: Global Public Opinion on International Affairs, 1 July 2009.
39. Scott Shane, "C.I.A. to Expand Use of Drones in Pakistan," *New York Times*, 3 December 2009.

40. George Orwell, *Nineteen Eighty-Four* (New York: Plume, 1983), p. 27.
41. Jawed Naqvi, "Singh Sees 'Vital Interest' in Peace with Pakistan," *Dawn*, 9 June 2009.
42. Ravi Nessman, "Ambitious India Now World's Largest Arms Importer," Associated Press, 13 March 2011.
43. Yossi Melman, "Media Allege Corruption in Massive Israel-India Arms Deal," *Ha'aretz* (Tel Aviv), 29 March 2009.
44. Prafulla Marpakwar, "Security Issues: City Team to Take Tips from Israel," *Times of India*, 11 July 2009. See also "Spy Drones to Be Deployed on Tamil Nadu Coast on Wednesday," *Times of India*, 10 April 2012.
45. Jane Hunter, *Israeli Foreign Policy: South Africa and Central America* (Boston: South End Press, 1987).
46. Chidanand Rajghatta, "Israeli Teams Training Forces in Kashmir: Jane's," *Times of India*, 16 August 2001. "Israelis Trained Kurds: BBC," *Dawn*, 21 September 2006. See also Benjamin Beit-Hallahmi, *The Israeli Connection: Who Israel Arms and Why* (New York: Pantheon Books, 1987).
47. For background, see Noam Chomsky, *Fateful Triangle: The United States, Israel, and the Palestinians*, updated ed. (Cambridge, MA: South End Press, 1999), p. 26.
48. Glen Carey, "Chinese Imports of Saudi Oil Will Rise 19% This Year to 50 Million Tons," Bloomberg News, 29 September 2010.
49. Kalbe Ali, "China Agrees to Run Gwadar Port," *Dawn*, 22 May 2011.
50. Associated Press, "Brazil Sets Trade Records, Due to Chinese Demand," 2 January 2012. The report notes, "In 2009 China surpassed the U.S. as Brazil's biggest trading partner."
51. Arundhati Roy, "Can We Leave the Bauxite in the Mountain?" Harvard University, Cambridge, Massachusetts, 1 April 2010.
52. Noam Chomsky, "When Elites Fail, and What We Should Do About It," First Unitarian Church, Portland, Oregon, 2 October 2009.
53. For a useful history, see Irving Bernstein, *The Lean Years: A History of the American Worker, 1920–1933*, updated ed. (Chicago: Haymarket Books, 2010), and *The Turbulent Years: A History of the American Worker, 1933–1941*, updated ed. (Chicago: Haymarket Books, 2010).
54. David Montgomery, *The Fall of the House of Labor: The Workplace, the State, and American Labor Activism, 1865–1925* (Cambridge: Cambridge University Press, 1987).
55. Douglas A. Fraser, Resignation Letter to the Labor–Management

Group (19 July 1978), reprinted in *Voices of a People's History of the United States*, ed. Howard Zinn and Anthony Arnove, 2nd ed. (New York: Seven Stories Press, 2010), pp. 529–33.

56. Noam Chomsky, "Closing Plenary: Rekindling the Radical Imagination," Left Forum, New York, New York, 21 March 2010. See also "Internet Note Posted by Man Linked to Plane Crash," *Austin Statesman*, 18 February 2010.
57. Michael Brick, "Man Crashes Plane into Texas I.R.S. Office," *New York Times*, 18 February 2010.
58. Chris Cillizza, "Vote Out the Entire Congress? You Bet," WashingtonPost.com, 6 September 2011.
59. See Daniel Guérin, *The Brown Plague: Travels in Late Weimar and Early Nazi Germany*, trans. Robert Schwartzwald (Durham, NC: Duke University Press, 1994). See also Peter Fritzsche, *Germans Into Nazis* (Cambridge, MA: Harvard University Press, 1998).
60. Scott Shane, "Conservatives Draw Blood from Acorn," *New York Times*, 15 September 2009.
61. "Exchange of Rail Know-How Between the United States and Spain," SpanishRailwayNews.com, 7 December 2011. See also Thomas Catan and David Gauthier-Villars, "Europe Listens for U.S. Train Whistle," *Wall Street Journal*, 29 May 2009.

2. CHAINS OF SUBMISSION AND SUBSERVIENCE

1. Stuart Ewen, *Captains of Consciousness: Advertising and the Social Roots of the Consumer Culture* (New York: Basic Books, 2001), p. 85.
2. For further discussion, see Noam Chomsky, *Necessary Illusions: Thought Control in Democratic Societies* (Boston: South End Press, 1989).
3. Ewen, *Captains of Consciousness*, p. 85.
4. Ben Arnoldy, "For Laid-Off IBM Workers, a Job in India?" *Christian Science Monitor*, 26 March 2009.
5. See Diane Ravitch, *The Death and Life of the Great American School System: How Testing and Choice Are Undermining Education* (New York: Basic Books, 2010).
6. Steven Greenhouse, "Union Membership in U.S. Fell to a 70-Year Low Last Year," *New York Times*, 21 January 2011.
7. Ross Eisenbrey, "Workers Want Unions Now More than Ever," Economic Policy Institute *Snapshot*, 28 February 2007, and Richard B. Freeman, *Do Workers Still Want Unions? More Than Ever* (Washington, DC: Economic Policy Institute, 2007).

8. Kate Bronfenbrenner, "A War Against Workers Who Organize," *Washington Post*, 3 June 2009.
9. Greenhouse, "Union Membership in U.S. Fell to a 70-Year Low Last Year."
10. See *Wisconsin Uprising: Labor Fights Back*, ed. Michael D. Yates (New York: Monthly Review Press, 2012).
11. Larry M. Bartels, "Inequalities," *New York Times Magazine*, 27 April 2008.
12. CNN, "Not Such a Lame-Duck Session: What Congress Passed, Obama Signed in Week," 23 December 2010.
13. Peter Baker, "With New Tax Bill, a Turning Point for the President," *New York Times*, 17 December 2010. Paul Sullivan, "Estate Tax Will Return Next Year, but Few Will Pay It," *New York Times*, 17 December 2010.
14. Peter Baker and Jackie Calmes, "Amid Deficit Fears, Obama Freezes Pay," *New York Times*, 29 November 2010.
15. Susanne Craig and Eric Dash, "Study Points to Windfall for Goldman Partners," *New York Times*, 18 January 2011.
16. Noam Chomsky, "Human Intelligence and the Environment," *International Socialist Review*, no. 76 (March–April 2011).

3. UPRISINGS

1. Eileen Byrne, "Death of a Street Seller That Set Off an Uprising," *Financial Times* (London), 16 January 2011.
2. For useful background, see Stephen Franklin, "In Egypt, Arab World's 'Largest Social Movement' Gains Steam Among Workers," *In These Times*, 28 June 2010.
3. John Thorne, "Tent City Is 'A Call for Independence,'" *The National* (Abu Dhabi), 8 November 2010.
4. For analysis, see Stephen Zunes and Jacob Mundy, *Western Sahara: War, Nationalism, and Conflict Irresolution* (Syracuse, NY: Syracuse University Press, 2010).
5. Rakesh Kochhar, Richard Fry, and Paul Taylor, "Wealth Gaps Rise to Record Highs Between Whites, Blacks, Hispanics," Pew Research Center, Washington, DC, 26 July 2011.
6. See "Median Net Worth of Households, 2005 and 2009," in Kochhar, Fry, and Taylor, "Wealth Gaps Rise to Record Highs."
7. Joel Beinin, "Egypt's Workers Rise Up," *Nation*, March 7–14, 2011. See also Amy Goodman's interview with Joel Beinin, "Striking

Egyptian Workers Fuel the Uprising After 10 Years of Labor Organizing," *Democracy Now!*, 10 February 2011.

8. "Tunisia's Islamist Ennahda Party Wins Historic Poll," BBC News Africa, 27 October 2011.
9. Borzou Daragahi, "Call for Probe into Libyan Civilian Deaths," *Financial Times* (London), 14 May 2012.
10. Agence France-Presse, "Diplomacy Takes Centre Stage in Libyan Conflict," 10 April 2011.
11. Staff report, "BRICS Leaders Call for Diplomatic Solution to Libya Crisis," *Nation* (Pakistan), 14 April 2011. See also Jo Ling Kent, "Leaders at BRICS Summit Speak Out Against Airstrikes in Libya," CNN, 14 April 2011; and Hugh Roberts, "Who Said Gaddafi Had to Go?," *London Review of Books*, 17 November 2011.
12. Eric Westervelt, "NATO's Intervention in Libya: A New Model?" National Public Radio, *Morning Edition*, 12 September 2011.
13. Abby Phillip, "Turkey Not Game to Back NATO," *Politico*, 28 March 2011.
14. Donald Macintyre, "Arab Support Wavers as Second Night of Bombing Begins," *Independent* (London), 21 March 2011.
15. "African Union Offers Truce Plan to Libyan Rebels," BBC News Africa, 11 April 2011.
16. Aijaz Ahmad, "Libya Recolonised," *Frontline* (India) 28, no. 3 (5–18 November 2011).
17. Eric Schmitt, "U.S. 'Gravely Concerned' over Violence in Libya," *New York Times*, 20 February 2011. Barack Obama, "Remarks by the President on the Situation in Libya," Office of the Press Secretary, 18 March 2011.
18. Tariq Ali, "Libya Is Another Case of Selective Vigilantism by the West," *Guardian* (London), 29 March 2011.
19. Toby Matthiesen, "Saudi Arabian Security Forces Quell 'Day of Rage' Protests," *Guardian* (London), 11 March 2011. See also Toby Matthiesen, "Saudi Arabia: The Middle East's Most Under-Reported Conflict," *Guardian* (London), 23 January 2012.
20. Ethan Bronner and Michael Slackman, "Saudi Troops Enter Bahrain to Help Put Down Unrest," *New York Times*, 14 March 2011.
21. Eyder Peralta, "Symbol of Uprising Is Destroyed in Bahrain," National Public Radio, *The Two Way* blog, 18 March 2011.
22. Neela Banerjee and David S. Cloud, "Medical Workers Caught in Bahrain Security Crackdown," *Los Angeles Times*, 21 March 2011.

23. Gerald F. Seib, "Pivotal Moment for America," *Wall Street Journal*, 12 February 2011.
24. Lloyd C. Gardner, *Three Kings: The Rise of an American Empire in the Middle East After World War II* (New York: New Press, 2009), p. 96.
25. Andre England and Sylvia Pfeifer, "Iraq's Proven Oil Reserves Soar by a Quarter," *Financial Times* (London), 4 October 2010. See also Christopher Helman, "The World's Biggest Oil Reserves," *Forbes*, 21 January 2010.
26. Suzanne Goldenberg, "Bush Commits Troops to Iraq for the Long Term," *Guardian* (London), 26 November 2007. See also Guy Raz, "Long-Term Pact with Iraq Raises Questions," National Public Radio, *Morning Edition*, 24 January 2008. For further analysis, see Noam Chomsky, *Hopes and Prospects* (Chicago: Haymarket Books, 2010).
27. Charlie Savage, "Bush Asserts Authority to Bypass Defense Act," *Boston Globe*, 30 January 2008.
28. GOP Debate, Myrtle Beach Convention Center, Myrtle Beach, South Carolina, 16 January 2012.
29. William S. Cohen, *Report of the Quadrennial Defense Review* (Washington, DC: Department of Defense, May 1997), p. 8.
30. See, for example, Glenn Greenwald, "Killing of Bin Laden: What Are the Consequences?" *Salon*, 2 May 2011.
31. Matthew Yglesias, "International Law Is Made by Powerful States," ThinkProgress.org, 13 May 2011. See also Matthew Yglesias, "Killing Osama Bin Laden Is Legal," ThinkProgress.org, 5 May 2011.
32. "Is America Over?" *Foreign Affairs* 6, no. 4 (November–December 2011).
33. Scott Shane, "Balancing U.S. Policy on an Ally in Transition," *New York Times*, 20 November 2011.
34. For discussion and references, see Noam Chomsky, *Turning the Tide: U.S. Intervention in Central America and the Struggle for Peace*, expanded ed. (Boston: South End Press, 1999), p. 66.
35. Thomas Carothers, *Critical Mission: Essays on Democracy Promotion* (Washington, DC: Carnegie Endowment for International Peace, 2004), pp. 7, 42.
36. Anthony Shadid, "At Mubarak Trial, Stark Image of Humbled Power," *New York Times*, 3 August 2011.
37. Senate Resolution 534, 109th Cong., 2d sess., 18 July 2006. For background, see Stephen Zunes, "Congress and the Israeli Attack on Lebanon: A Critical Reading," *Foreign Policy in Focus*, 22 July 2006.

38. Hans J. Morgenthau, *The Purpose of American Politics* (New York: Alfred A. Knopf, 1960).
39. Ibid., p. 7.
40. Ibid.

4. DOMESTIC DISTURBANCES

1. Press Release, "Occupy the Dream Rolls Out National Steering Committee to Join the Occupy Wall Street Movement," Occupy the Dream, Washington, DC, 3 January 2012.
2. Charlie Savage, "Obama Drops Veto Threat Over Military Authorization Bill After Revisions," *New York Times*, 14 December 2011.
3. For a detailed analysis, see Glenn Greenwald, "Three Myths about the Detention Bill," *Salon*, 16 December 2011.
4. Supreme Court of the United States, *Holder*, Attorney General et al. v. Humanitarian Law Project et al., Washington, DC, no. 08–1498. Argued 23 February 2010. Decided 21 June 2010.
5. U.S. Department of State, Bureau of Counterterrorism, Foreign Terrorist Organizations, 27 January 2012. Available at http://www.state.gov/j/ct/rls/other/des/123085.htm.
6. For analysis, see David Cole, "Advocacy Is Not a Gun," *New York Times*, *Room for Debate* blog, 21 June 2010.
7. "Mandela Taken Off US Terror List," BBC News, 1 July 2008.
8. David B. Ottaway, "Iraq Gives Haven to Key Terrorist," *Washington Post*, 9 November 1982.
9. "Mandela Taken Off US Terror List."
10. Robert Pear, "U.S. Report Stirs Furor in South Africa," *New York Times*, 13 January 1989.
11. Andy Grimm and Cynthia Dizikes, "FBI Raids Anti-war Activists' Homes," *Chicago Tribune*, 24 September 2010.
12. Glenn Greenwald, "The Omar Khadr Travesty," *Salon*, 11 August 2010.
13. Charlie Savage, "Delays Keep Former Qaeda Child Soldier at Guantánamo, Despite Plea Deal," *New York Times*, 24 March 2012.
14. Lawyers Rights Watch Canada, "Canada in Breach of Human Rights Obligations in Omar Khadr Case," Vancouver, British Columbia, 16 May 2012.
15. GOP Debate, Myrtle Beach Convention Center, Myrtle Beach, South Carolina, 16 January 2012.
16. Ibid.

17. Jeffrey M. Jones, "Unemployment Re-Emerges as Most Important Problem in the U.S.," Gallup, 15 September 2011.
18. Immanuel Wallerstein, interview with Sophie Shevardnadze, *Russia Today*, 4 October 2011.
19. Martin Wolf, "The Big Question Raised by Anti-Capitalist Protests," *Financial Times* (London), 28 October 2011.
20. See also Richard Wolff, *Democracy at Work* (Chicago: Haymarket Books, 2012).
21. Howard Zinn, "A Chorus Against War," *The Progressive* 67, no. 3 (March 2003), pp. 19–21.
22. Howard Zinn, "Operation Enduring War," *The Progressive* 66, no. 3 (March 2002), pp. 12–13.
23. David Hume, "Of the First Principles of Government," in *Selected Essays*, ed. Stephen Copley and Andrew Edgar (New York: Oxford University Press, 1996), p. 24.
24. Edward Bernays, *Propaganda* (Brooklyn: Ig Publishing, 2005), p. 127.
25. Clinton Rossiter and James Lare, *The Essential Lippmann: A Political Philosophy for Liberal Democracy* (Cambridge, MA: Harvard University Press, 1965), p. 91.
26. David Brooks, "Midlife Crisis Economics," *New York Times*, 26 December 2011. Elizabeth Mendes, "In U.S., Fear of Big Government at Near-Record Level," Gallup, 12 December 2011.
27. See Pew Research Center for the People and the Press, "Question Wording," no date, available at http://www.people-press.org/methodology/questionnaire-design/question-wording/.
28. Eva Bertram, "Democratic Divisions in the 1960s and the Road to Welfare Reform," *Political Science Quarterly* 126, no. 4 (Winter 2011–12), pp. 579–610.
29. Barbara Vobejda, "Clinton Signs Welfare Bill Amid Division," *Washington Post*, 23 August 1996.
30. Aristotle, *The Politics and the Constitution of Athens* (Cambridge: Cambridge University Press, 1996), p. 75.
31. James Madison, The Federalist No. 10 ("The Utility of the Union as a Safeguard Against Domestic Faction and Insurrection [Continued]"), *Daily Advertiser*, 22 November 1787.

5. UNCONVENTIONAL WISDOM

1. See Dean Baker, "Faith-Based Economics at the European Central Bank," *Guardian* (London), 11 April 2012.

2. Dakin Campbell, "U.S. Banks Face Contagion Risk from Europe Debt," Bloomberg News, 17 November 2011.
3. Michael Mackenzie, Dan McCrum, and Lindsay Whipp, "US Treasuries: Surprisingly Sturdy," *Financial Times* (London), 15 December 2011.
4. Richard Wolff, "Occupy Wall Street and the Economic Crisis," New York, New York, 20 November 2011 (*Alternative Radio*, no. WOLR004). See also Richard Wolff and David Barsamian, *Occupy the Economy: Challenging Capitalism* (San Francisco: City Lights Books, 2012).
5. Dan Bilefsky and Sebnem Arsu, "Charges Against Journalists Dim the Democratic Glow in Turkey," *New York Times*, 4 January 2012.
6. For a detailed discussion, see Kerim Yildiz, *The Kurds in Turkey: EU Accession and Human Rights* (London: Pluto Books, 2005).
7. Tamar Gabelnick, William D. Hartung, and Jennifer Washburn, with Michelle Ciarrocca, *Arming Repression: U.S. Arms Sales to Turkey During the Clinton Administration* (New York and Washington, DC: World Policy Institute and Federation of Atomic Scientists, October 1999). See Table I: "Total Dollar Value of U.S. Arms Deliveries to Turkey Through the Direct Commercial Sales (DCS) and Foreign Military Sales (FMS) Programs from FY 1950 to 1998."
8. Jim Lobe, "Erdogan Most Popular Leader by Far Among Arabs," Inter Press Service, 21 November 2011. James Zogby, *Arab Attitudes, 2011* (Washington, DC: Arab American Institute Foundation, 2011), p. 1.
9. Souren Melikian, "Turkey Reawakening to Its Vast Iranian Ties," *New York Times*, 23 April 2010.
10. David E. Sanger and Michael Slackman, "U.S. Is Skeptical on Iranian Deal for Nuclear Fuel," *New York Times*, 17 May 2010. See also Mark Landler, "At the U.N., Turkey Asserts Itself in Prominent Ways," *New York Times*, 22 September 2010.
11. Alexei Barrionuevo, "Obama Writes to Brazil's Leader About Iran," *New York Times*, 24 November 2009.
12. Phillip, "Turkey Not Game to Back NATO," 28 March 2011.
13. Sebnem Arsu, "Turkey Lashes Out over French Bill About Genocide," *New York Times*, 23 December 2011.
14. "EU to Tell Turkey to Shape Up," *New York Times*, 4 October 2008.
15. See *Midnight on the Mavi Marmara: The Attack on the Gaza Freedom Flotilla and How It Changed the Course of the Israel/Palestine Conflict*, ed. Moustafa Bayoumi (Chicago: Haymarket Books, 2010).
16. Megan K. Stack, "Israel Flotilla Raid Deals a Blow to Ties with Turkey," *Los Angeles Times*, 31 May 2010.

17. For discussion, see Paul Street, "Obama-Gaza: No Surprise," ZCommunications.org, 4 January 2009.
18. Katrin Bennhold, "Leaders of Turkey and Israel Clash at Davos Panel," *New York Times*, 29 January 2009.
19. Talila Nesher, "Israeli MKs to Discuss Recognizing Turkey's Armenian Genocide," *Ha'aretz* (Tel Aviv), 26 December 2011.
20. Peter Balakian, "State of Denial," *Tablet*, 19 October 2010. See also Israel W. Charney, "A Moral Israel Must Recognize the Armenian Genocide," *Jerusalem Post Magazine*, 24 January 2012.
21. Raphael Ahren, "Genocide Expert Calls on Israel to Put Armenian Suffering Before Politics," *Ha'aretz* (Tel Aviv), 22 July 2011.
22. Israel W. Charny, "Fighting for Israel's Recognition of the Armenian Genocide," *Genocide Prevention Now*, no. 7 (Summer 2011).
23. Associated Press, "Israel Snubs Turkish Ambassador in Public," 12 January 2010.
24. Sabrina Tavernise, "Raid Jeopardizes Turkey Relations," *New York Times*, 31 May 2010. See also Marc Champion and Joshua Mitnick, "Turkey Expels Israeli Ambassador," *Wall Street Journal*, 3 September 2011.
25. For a detailed history, see Jonathan C. Randal, *After Such Knowledge, What Forgiveness?: My Encounters with Kurdistan* (Boulder, CO: Westview Press, 1998).
26. Kevin McKiernan, *Good Kurds, Bad Kurds* (Access Productions, 2001), 81 mins.
27. Kim Sengupta and Donald Macintyre, "Israel's Military Leaders Warn Against Iran Attack," *Independent* (London), 2 February 2012. Mark Mazzetti and Thom Shanker, "U.S. War Game Sees Perils of Israeli Strike Against Iran," *New York Times*, 19 March 2012. See also Elisabeth Bumiller, "Iran Raid Seen as a Huge Task for Israeli Jets," *New York Times*, 19 February 2012.
28. Karl Vick and Aaron J. Klein, "Who Assassinated an Iranian Nuclear Scientist? Israel Isn't Telling," *Time*, 13 January 2012. Indira A. R. Lakshmanan, "Iran Is Seen Suffering Crippling Effect of Sanctions on Oil Trade, Banking," Bloomberg News, 29 February 2012. Shirzad Bozorgmehr and Moni Basu, "Sanctions Take Toll on Ordinary Iranians," CNN, 23 January 2012.
29. Klaus Naumann et al., *Towards a Grand Strategy for an Uncertain World: Renewing the Transatlantic Partnership* (Lunteren, The Netherlands: Noaber Foundation, 2007), p. 27.

30. Ibid., p. 97. See also John J. Kruzel, "Gates Discusses New Nuclear Posture, U.S. Relations with Karzai," American Forces Press Service, 11 April 2010.
31. Orwell, *Nineteen Eighty-Four*, p. 39.
32. Noam Chomsky, "The Torture Memos and Historical Amnesia," *Nation*, 1 June 2009, no. 40, p. 179.
33. Anatol Lieven, "Afghanistan: The Best Way to Peace," *New York Review of Books*, 9 February 2012.
34. Jane Perlez, "Pakistanis Continue to Reject U.S. Partnership," *New York Times*, 30 September 2009. See also Pew Global Attitudes Project, "Public Opinion in Pakistan: Concern About Extremist Threat Slips: America's Image Remains Poor," 29 July 2010.
35. Scott Shane, "Drone Strike Kills Qaeda Operative in Pakistan, U.S. Says," *New York Times*, 19 January 2012.
36. "US Embassy Cables: 'Reviewing Our Afghanistan-Pakistan Strategy,'" *Guardian* (London), 30 November 2010.

6. MENTAL SLAVERY

1. Bob Marley and the Wailers, "Redemption Song," *Uprising* (Tuff Gong/Island, 1980).
2. Matthew Creamer, "Obama Wins! . . . Ad Age's Marketer of the Year," *Advertising Age*, 17 October 2008.
3. John Quelch, "How Better Marketing Elected Barack Obama," *Harvard Business Review, HRB Blog* Network, 5 November 2008. See also Andrew Edgecliffe-Johnson, "Bush Set to Be Knocked Off His CEO Pedestal," *Financial Times*, 25 November 2008.
4. E. L. Doctorow, *Ragtime* (New York: Plume, 1997).
5. Mohammed Hanif, *A Case of Exploding Mangoes* (New York: Alfred A. Knopf, 2008).
6. James Rainey, "Wikipedia to Go Offline to Protest Anti-Piracy Legislation," *Los Angeles Times*, 17 January 2012.
7. Jonathan Weisman, "In Fight over Piracy Bills, New Economy Rises Against Old," *New York Times*, 18 January 2012.
8. Dean Baker, *Financing Drug Research: What Are the Issues?* (Washington, DC: Center for Economic and Policy Research, September 2004).
9. Dean Baker, "The Surefire Way to End Online Piracy: End Copyright," Truthout.org, 23 January 2012.
10. "A Selection from the Cache of Diplomatic Dispatches," "Diplomacy: Analyzing a Coup in Honduras," *New York Times*, 19 June 2011.

11. Robert Naiman, "WikiLeaks Honduras: State Dept. Busted on Support of Coup," CommonDreams.org, 29 November 2010.
12. David E. Sanger, James Glanz, and Jo Becker, "Around the World, Distress Over Iran," *New York Times*, 28 November 2010.
13. Jacob Heilbrunn, "Are the WikiLeaks Actually an American Plot?" *National Interest*, 29 November 2010.
14. *2010 Arab Public Opinion Survey* (Washington, DC: Zogby International/Brookings Institution, 2010). The figure is from among Egyptians "who believe Iran has peaceful goals," and is 69 percent for Egyptians "who believe Iran seeks nuclear weapons."
15. In response to the question, "Name two countries that you think pose the biggest threat to you," Israel received 88 percent, the United States 77 percent, and Iran 9 percent among those aged thirty-six and over and 11 percent among those thirty-six and under. *2010 Arab Public Opinion Survey*.
16. Ian Black, "WikiLeaks Cables: Tunisia Blocks Site Reporting 'Hatred' of First Lady," *Guardian* (London), 7 December 2010. Ian Black, "Profile: Zine al-Abidine Ben Ali," *Guardian* (London), 14 January 2011. See also Amy Davidson, "Tunisia and WikiLeaks," *New Yorker, Close Read* blog, 14 January 2011.
17. Steven Erlanger, "French Foreign Minister Urged to Resign," *New York Times*, 3 February 2011.
18. Charlie Savage, "Soldier Faces 22 New WikiLeaks Charges," *New York Times*, 2 March 2011.
19. Scott Shane, "Court Martial Recommended in WikiLeaks Case," *New York Times*, 12 January 2012.
20. Stephanie Condon, "Obama Says Bradley Manning 'Broke the Law,'" CBSNews.com, 22 April 2011.
21. Mark Mazzetti, Eric Schmitt, and Robert F. Worth, "Two-Year Manhunt Led to Killing of Awlaki in Yemen," *New York Times*, 30 September 2011.
22. President Barack Obama, "President Obama's Statement on the Memos," *New York Times*, 16 April 2009. See also Mark Mazzetti and Scott Shane, "Interrogation Memos Detail Harsh Tactics by the C.I.A.," *New York Times*, 16 April 2009.
23. Noam Chomsky, "If the Nuremberg Laws Were Applied . . ." Chomsky.info, circa 1990, available at http://www.chomsky.info/talks/1990----.htm.
24. Principles of International Law Recognized in the Charter of the Nuremberg Tribunal and in the Judgment of the Tribunal, 1950.

25. See Telford Taylor, *Nuremberg and Vietnam: An American Tragedy* (Chicago: Quadrangle, 1970), p. 39. See also Telford Taylor, *The Anatomy of the Nuremberg Trials: A Personal Memoir* (New York: Alfred A. Knopf, 1992).
26. Taylor, *The Anatomy of the Nuremberg Trials*, p. 567. See also Taylor, *Nuremberg and Vietnam*, pp. 37 and 86.
27. Alex Pareene, "Our Militarized Police Forces," *Salon*, 8 November 2011.
28. Noam Chomsky, "Who Owns the World?" TomDispatch.com, 21 April 2011.
29. Fiona Harvey, "World Headed for Irreversible Climate Change in Five Years, IEA Warns," *Guardian* (London), 9 November 2011.
30. Ibid.
31. Ibid. See also Andrew Revkin, "High Odds of Hot Times," *New York Times, Dot Earth* blog, 20 May 2009. See also David Chandler, "Climate Change Odds Much Worse than Thought," *MIT News*, 19 May 2009.
32. Edward Luce, "America Is Entering a New Age of Plenty," *Financial Times* (London), 20 November 2011.
33. Naomi Klein, "Capitalism vs. the Climate," *Nation*, 28 November 2011.
34. Clifford Krauss and Jad Mouawad, "Oil Industry Backs Protests of Emissions Bill," *New York Times*, 18 August 2009.
35. Davis Asman, Interview with Ron Paul, Fox Business, 4 November 2009.

7. LEARNING HOW TO DISCOVER

1. William James, *The Principles of Psychology*, vol. 1 (New York: Henry Holt, 1918), p. 488.
2. Bill Keller, "Diplomats and Dissidents," *New York Times*, 13 May 2012.
3. *The Cambridge History of the Cold War*, ed. Melvyn P. Leffler and Odd Arne Westad, 3 vols. (Cambridge: Cambridge University Press, 2012). See particularly John H. Coatsworth, "The Cold War in Central America, 1975–1991," *The Cambridge History of the Cold War*, vol. 3, p. 221.
4. For further discussion, see Noam Chomsky, *Hopes and Prospects* (Chicago: Haymarket Books, 2010).
5. Noam Chomsky, *Chomsky on Democracy and Education*, ed. Carlos P. Otero (New York: RoutledgeFalmer, 2003), p. 34.

6. Ralph Waldo Emerson, *The Works of Ralph Waldo Emerson*, vol. 2 (London: Macmillan, 1883), p. 525.
7. For discussion, see Noam Chomsky, *Current Issues in Linguistic Theory* (New York: Mouton de Gruyter, 1964). See also Noam Chomsky, *Cartesian Linguistics: A Chapter in the History of Rationalist Thought* (Cambridge: Cambridge University Press, 2009).
8. Michel Crozier, Samuel P. Huntington, and Joji Watanuki, *The Crisis of Democracy: Report on the Governability of Democracies to the Trilateral Commission* (New York: New York University Press, 1975), p. 113.
9. Lewis F. Powell Jr., Confidential Memorandum: Attack on American Free Enterprise System, August 23, 1971, available at http://www.greenpeace.org/usa/en/campaigns/global-warming-and-energy/polluterwatch/The-Lewis-Powell-Memo/.
10. National Security Council Report 68: "United States Objectives and Programs for National Security," 14 April 1950, available at www.fas.org/irp/offdocs/nsc-hst/nsc-68.htm.
11. Crozier, Huntington, and Watanuki, *The Crisis of Democracy*, p. 162.
12. Andrew Martin and Andrew W. Lehren, "A Generation Hobbled by the Soaring Cost of College," *New York Times*, 12 May 2012. Janet Lorin, "Student-Loan Debt Reaches Record $1 Trillion, Report Says," Bloomberg News, 22 March 2012.
13. Ron Lieber, "Student Debt and a Push for Fairness," *New York Times*, 4 June 2010.
14. On racism under the GI bill, see Ira Katznelson, *When Affirmative Action Was White: An Untold History of Racial Inequality in Twentieth-Century America* (New York: W. W. Norton, 2005), p. 114.
15. Paul de la Garza, "Mexico Students Strike over Higher Fees," *Chicago Tribune*, 20 May 1999. Julia Preston, "University Officials Yield to Student Strike in Mexico," *New York Times*, 8 June 1999.
16. Tim Walker, "In High-Performing Countries, Education Reform Is a Two-Way Street," *NEA Today*, 31 March 2011.
17. Diane Ravitch, "What Can We Learn from Finland?" *Education Week*, 11 October 2011.
18. See, among others, Bruce Alberts, "Considering Science Education," *Science*, 21 March 2008; "Making a Science of Education," *Science*, 2 January 2009; "Redefining Science Education," *Science*, 23 January 2009; "Prioritizing Science Education," *Science*, 23 April 2010; "An Education That Inspires," *Science*, 22 October 2010; and "Teaching Real Science," *Science*, 27 January 2012.
19. Alberts, "Teaching Real Science."

20. Dean Baker and Mark Weisbrot, *Social Security: The Phony Crisis* (Chicago: University of Chicago Press, 2000).
21. Michael Muskal, "Support at GOP Debate for Letting the Uninsured Die," *Los Angeles Times*, 13 September 2011.
22. Kate Nocera, "Rand Paul: 'Right to Health Care' Is Slavery," *Politico*, 11 May 2011.
23. *Survey of Young Americans' Attitudes Toward Politics and Public Service*, 21st ed. (Cambridge, MA: Harvard University Institute of Politics, 24 April 2012).
24. Ibid. Executive Summary, p. 18.
25. Ibid.

8. ARISTOCRATS AND DEMOCRATS

1. Michael P. Schmidt, "President Speaks Out on Guard Investigation," *New York Times*, 15 April 2012. Noam Chomsky, "Cartagena Beyond the Secret Service Scandal," *In These Times*, 2 May 2012.
2. Jennifer Ditchburn, "Emboldened Latin America Parts Ways with Canada, U.S. on Cuba and Drugs," *Toronto Star*, 14 April 2012.
3. Daniel Wallis and Andrew Cawthorne, "Lively Chavez Hosts Latin American Peers, Snubs U.S.," Reuters, 3 December 2011.
4. Evan Perez, "Mexican Guns Tied to U.S.," *Wall Street Journal*, 10 June 2011.
5. Chris McGreal, "How Mexico's Drug Cartels Profit from Flow of Guns Across the Border," *Guardian* (London), 8 December 2011. See also Richard A. Serrano, "ATF Fast and Furious Guns Turned up in El Paso," *Los Angeles Times*, 29 September 2011.
6. Tim Murphy, "Rand Paul Backs Fringe UN Gun Conspiracy," *Mother Jones*, 6 October 2011.
7. Nick Hopkins, "Minister Calls for Support for Tough New Arms Trade Treaty," *Guardian* (London), 16 May 2012.
8. George Parker, "UK to Push for UN Arms Trade Treaty," *Financial Times* (London), 16 May 2012. For detailed analysis, see *Small Arms Survey 2011: States of Security* (Cambridge: Cambridge University Press, 2011).
9. Theophilos Argitis and Jeremy Van Loon, "Obama's Keystone Denial Prompts Canada to Look to China Sales," Bloomberg News, 19 January 2012.
10. Barack Obama, "President Obama's State of the Union Address," *New York Times*, 25 January 2012.

11. Roy, *Field Notes on Democracy*. See also Arundhati Roy, *Walking the Comrades* (New York: Penguin Books, 2011).
12. Josh Fox, *Gasland* (Docurama Films, 2010), 107 mins.
13. Judy Battista, "Vikings Will Remain in Minnesota," *New York Times*, 10 May 2012.
14. Steven Salzberg, "University of Florida Eliminates Computer Science Department, Increases Athletic Budgets. Hmm.," *Forbes*, 22 April 2012.
15. Dave Zirin, "No Class: College Football Coach Salaries Rose 35 Percent Last Year," *Nation*, 21 January 2012.
16. Kristen A. Graham, "Phila[delphia] School District Plan Includes Restructuring and School Closings," *Philadelphia Inquirer*, 24 April 2012.
17. "California State U[niversity] Faculty Members Give Green Light to Rolling Strikes," *Chronicle of Higher Education*, 2 May 2012.
18. Nanette Asimov, "Cal State to Close Door on Spring 2013 Enrollment," *San Francisco Chronicle*, 20 March 2012.
19. Benjamin Ginsberg, *The Fall of the Faculty: The Rise of the All-Administrative University and Why It Matters* (Oxford, UK: Oxford University Press, 2011).
20. Josh Bivens, *Failure by Design: The Story Behind America's Broken Economy* (Ithaca, NY: Cornell University Press, 2011).
21. Brian Blackstone, Matthew Karnitschnig, and Robert Thomson, "Europe's Banker Talks Tough," *Wall Street Journal*, 24 February 2012.
22. Scott DeCarlo, "The World's 25 Most Valuable Companies: Apple Is Now on Top," *Forbes*, 11 August 2011. David Barboza, "After Suicides, Scrutiny of China's Grim Factories," *New York Times*, 6 June 2010.
23. Charles Duhigg and David Kocieniewski, "How Apple Sidesteps Billions in Taxes," *New York Times*, 28 April 2012.
24. Robert Reich, "The Answer Isn't Socialism; It's Capitalism That Better Spreads the Benefits of the Productivity Revolution," RobertReich.org, 6 May 2012, available at http://robertreich.org/post/22542609387.
25. See the website for International Organization for a Participatory Society (IOPS) at http://www.iopsociety.org/.
26. William Rogers, "USW and Mondragon Announce New Worker Co-op Plan," *Left Labor Reporter*, 2 April 2012.

27. Mikhail Bakunin, letter to Sergey Nechayev, 2 June 1870.
28. Noam Chomsky, "Democracy and Education," Loyola University, Chicago, Illinois, 19 October 1994 (*Alternative Radio*, no. CHON108).
29. Charles Sellers, *The Market Revolution: Jacksonian America, 1815–1846* (New York: Oxford University Press, 1991), p. 269.
30. *The Jeffersonian Cyclopedia: A Comprehensive Collection of the Views of Thomas Jefferson*, ed. John P. Foley (New York: Funk & Wagnalls Company, 1900), p. 49.
31. Ibid.
32. *Bakunin on Anarchism*, ed. Sam Dolgoff (Montréal: Black Rose Books, 2002), p. 330.
33. Daniel Guérin, *Jeunesse du socialisme libertaire: essais* (Paris: Librairie Marcel Rivière et Cie, 1959), p. 119.
34. Supreme Court of the United States, *Citizens United v. Federal Election Commission*, Washington, DC, no. 8-205. Argued 24 March 2009. Reargued 9 September 2009. Decided 21 January 2010. Michael Bonanno, "Democracy Unlimited of Humboldt County Launches Move to Amend the Constitution," OpEdNews.com, 22 January 2010.
35. Jason Burke, "Bhopal Campaigners Condemn 'Insulting' Sentences over Disaster," *Guardian* (London), 7 June 2010.
36. Weisbrot and Watkins, "Recent Experiences with International Financial Markets."
37. Supreme Court of the United States, *Buckley v. Valeo*, Washington, DC, no. 75-436. Argued 10 November 1975. Decided 30 January 1976.
38. Burt Neuborne, "Why the ACLU Is Wrong About 'Citizens United,'" *Nation*, 9 April 2012.
39. Nicholas Sonfessore, "'Super PACs' Let Strategists Off the Leash," *New York Times*, 20 May 2012.
40. Karl Marx, "Theses on Feuerbach," in *Writings of the Young Marx on Philosophy and Society*, ed. Lloyd David Easton and Kurt H. Guddat (New York: Doubleday, 1967), p. 402.

ACKNOWLEDGMENTS

Special thanks to Anthony Arnove, Sara Bershtel, Sophie Siebert, and Bev Stohl. Excerpts of these interviews appeared in the *International Socialist Review* (www.isreview.org) and aired on KGNU and Alternative Radio.

INDEX

read more

NOAM CHOMSKY

WHO RULES THE WORLD?

'The closest thing in the English-speaking world to an intellectual superstar' ***Guardian***

As long as the general population is passive, apathetic, diverted to consumerism or hatred of the vulnerable, the powerful can do as they please and those who survive will be left to contemplate the outcome . . .

In the post-9/11 era, America's policy-makers have increasingly prioritised the pursuit of power, both military and economic, above all else – human rights, democracy, even security. Drawing on examples ranging from expanding drone assassination programs to civil war in Syria to the continued violence in Iraq, Iran, Afghanistan, Israel and Palestine, philosopher, political commentator and prolific activist Noam Chomsky offers unexpected and nuanced insights into the workings of imperial power in our increasingly chaotic planet.

NOAM CHOMSKY

HOW THE WORLD WORKS

'Chomsky has an authority granted by brilliance' *Sunday Times*

Divided into four sections and originally published in the US as individual short books which have collectively sold over half a million copies, *How the World Works* is a collection of speeches and interviews with Chomsky by David Barsamian, edited by Arthur Naiman. It includes 'What Uncle Sam Really Wants', about US foreign policy; 'The Prosperous Few and the Restless Many', about the new global economy, food, Third World 'economic miracles' and the roots of racism; 'Secrets, Lies and Democracy', about the US, the CIA, religious fundamentalism, global inequality and the coming eco-catastrophe; and 'The Common Good', about equality, freedom, the media, the myth of Third World debt and manufacturing dissent.

With exceptional clarity and power of argument, Noam Chomsky lays bare as no one else can the realities of contemporary geopolitics.

NOAM CHOMSKY

HEGEMONY OR SURVIVAL

'For anyone wanting to find out about the world we live in, there is one simple answer: read Noam Chomsky' ***New Statesman***

The United States is in the process of staking out not just the globe, but the last unarmed spot in our neighbourhood – the skies – as a militarized sphere of influence. Our earth and its skies are, for the Bush administration, the final frontiers of imperial control. In *Hegemony or Survival*, Noam Chomsky explains how we came to this moment, what kind of peril we find ourselves in, and why our rulers are willing to jeopardize the future of our species. In our era, Chomsky argues, empire is a recipe for an earthly wasteland.

From the world's foremost intellectual activist comes an irrefutable analysis of America's pursuit of total domination and the catastrophic consequences that are sure to follow.

NOAM CHOMSKY

FAILED STATES

'One of the greatest, most radical public thinkers of our time'
Arundhati Roy

The United States asserts the right to use military force against 'failed states' around the globe. But as Noam Chomsky argues in this devastating analysis, America shares features with many of the regimes it insists are failing and constitute a danger to their neighbours.

Offering a comprehensive and radical examination of America past and present, Chomsky shows how this lone superpower – which topples foreign governments, invades states that threaten its interests and imposes sanctions on regimes it opposes – has stretched its own democratic institutions to breaking point. And how an America in crisis places the world ever closer to the brink of nuclear and environmental disaster.

NOAM CHOMSKY

HOPES AND PROSPECTS

'One of our most valued political thinkers' ***Independent on Sunday***

In this urgent new book, Noam Chomsky surveys the threats and prospects of our early twenty-first century. Exploring challenges such as the growing gap between North and South, American exceptionalism (even under Obama), the fiascos of Iraq and Afghanistan, the Israeli assault on Gaza and the recent financial bailouts, he also sees hope for the future and a way to move forward – in the so-called democratic wave in Latin America and in the global solidarity movements which suggest 'real progress towards freedom and justice'.

Hopes and Prospects is essential reading for anyone who is concerned about the primary challenges still facing the human race and is wondering where to find a ray of hope.